This intriguing series encompasses exciting trends and discoveries in areas of human exploration and progress: astronomy, anthropology, biology, physics, geology, medicine, health, genetics, and evolution. Sometimes controversial, these timely volumes present stimulating new points of view about our universe... and ourselves. Among the titles:

Ancestral Voices: Language and the Evolution of Human Consciousness
Curtis G. Smith

The Human Body and Why It Works
Raymond L. Powis

Toward a New Brain: Evolution and the Human Mind
Stuart Litvak and A. Wayne Senzee

On Civilized Stars: The Search for Intelligent Life in Outer Space
Joseph F. Baugher

JOSEPH F. BAUGHER

ON CIVILIZED STARS

The Search for Intelligent Life in Outer Space

Prentice-Hall, Inc., Englewood Cliffs, New Jersey 07632

Library of Congress Cataloging in Publication Data

Baugher, Joseph F.
On civilized stars

(Frontiers of science)
"A Spectrum Book."
Bibliography: p.
Includes index.
1. Life on other planets. I. Title. II. Series.
QB54.B38 1985 574.999 84-18374
ISBN 0-13-634429-1
ISBN 0-13-634411-9 (pbk.)

TO JUDITH

A Spectrum Book. Printed in the United States of America.

10 9 8 7 6 5 4 3 2 1

Editorial/production supervision by Chris McMorrow
Manufacturing buyer: Frank Grieco
Cover design by Hal Siegel

ISBN 0-13-634429-1

ISBN 0-13-634411-9 {PBK.}

This book was set in Aster and Univers typefaces.

PRENTICE-HALL INTERNATIONAL, INC., *London*
PRENTICE-HALL OF AUSTRALIA PTY. LIMITED, *Sydney*
PRENTICE-HALL CANADA INC., *Toronto*
PRENTICE-HALL HISPANOAMERICANA, S.A., *Mexico*
PRENTICE-HALL OF INDIA PRIVATE LIMITED, *New Delhi*
PRENTICE-HALL OF JAPAN, INC., *Tokyo*
PRENTICE-HALL OF SOUTHEAST ASIA PTE. LTD., *Singapore*
WHITEHALL BOOKS LIMITED, *Wellington, New Zealand*
EDITORA PRENTICE-HALL DO BRASIL LTDA., *Rio de Janeiro*

CONTENTS

PREFACE

Approximately five billion years ago the Sun began to form out of diffuse gas and dust. There was nothing at first sight to distinguish it from myriads of other stars. Like all the others, the new Sun began to burn and emit light as it contracted. During the final stages of condensation, it happened to produce a large surrounding nebula of diffuse gas and dust. The particles and molecules of the nebula collided with each other and coalesced into successively larger and larger bodies. Eventually these bodies formed into large planets circling around the central star.

Some of the newly formed worlds were large gaseous bodies made up primarily of hydrogen. These planets differed very little in structure from the Sun itself. Other planets were small, rocky, airless worlds with surfaces that were scarred by the impact of giant meteorites. Still others were enveloped in dense, choking atmospheres of poisonous gases. Many were frozen wastelands completely covered with solid ice. All were indescribably hostile to living creatures.

One of these new worlds was different: It had a warm, pleasant climate with continents, oceans, running water, and blue skies. It was the third planet from the Sun, a place that came to be known as Earth. The oceans and the surface of Earth soon became teeming with life. Some of the creatures that evolved here had supple hands, endowing them with manual dexterity. They developed large brains that gave them intelligence. Human beings evolved, endowed with a passionate curiosity about their surroundings and a desire to understand their place in the cosmos. Their large brains and supple hands enabled them to escape from a total dependence on the capricious whims of nature; they could change their world to improve their lives and enhance their chance of survival. Human beings developed technology. They built cities, created religions, developed writing, produced science and mathematics. Human life and culture spread to every available niche on Earth. Human beings came to dominate the entire surface of the world within only a few thousand years of their

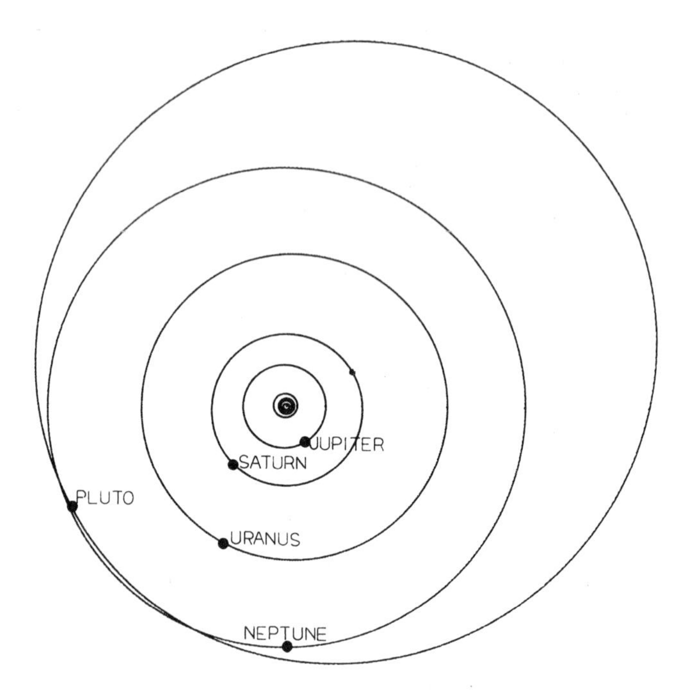

FIGURE A. The orbits of the planets in the outer solar system. The positions of the planets correspond to their locations relative to the Sun on January 1, 1985.

appearance. Now that humans are the undisputed masters of their planet, they are looking upward to the skies in search of new worlds to conquer.

When the history of the twentieth century is written in the far distant future, our era may be chiefly remembered as the time when humanity began its expansion into outer space. We have already begun a detailed reconnaissance of the solar system; our spacecraft have explored all the planets that were known to the ancients, and men have walked upon the surface of the Moon. A space transportation system is under development in the United States that promises to make operations in near-Earth orbit almost as routine as airline trips across the Atlantic Ocean. Teams of Soviet astronauts have been able to live and work in outer space for times of almost a year. Plans are currently being made for a permanent presence in outer space, with space stations containing thousands of inhabitants being seriously considered. Many of the planets and moons in our solar system may eventually be settled by human beings, with the surfaces of these worlds being purposefully altered to make them suitable for human habitation. Perhaps most of humanity will someday live in outer space, with Earth being set aside as a sort of wildlife preserve. The move into space is an event as significant in evolutionary history as the first movement of life onto dry land nearly four hundred million years ago.

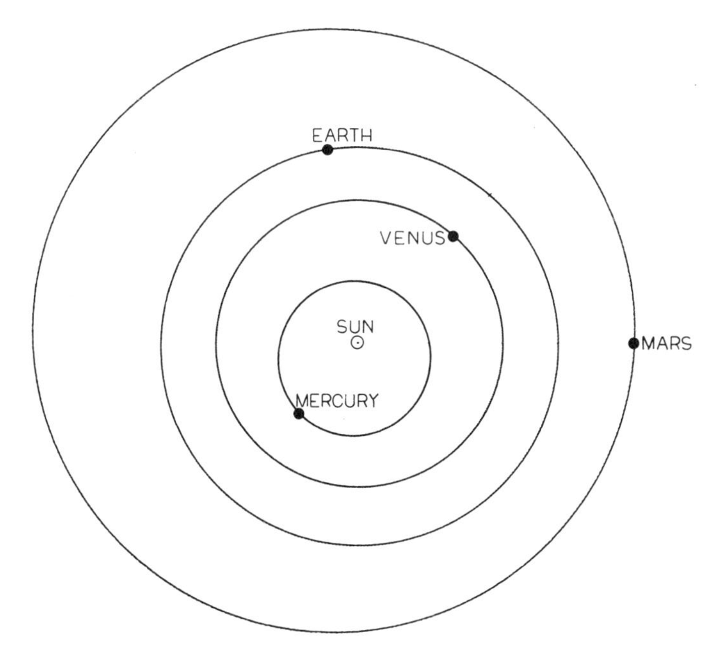

FIGURE B. The orbits of the inner four planets are shown in magnified view.

The drama of life and intelligence on Earth is a saga that has lasted for the past three and a half billion years and, we hope, shall continue into the indefinite future. Has this story been repeated countless numbers of times in many different variations on other worlds, or is the Earth somehow unique? Have other worlds evolved conscious, thinking beings who strive to understand the universe and themselves? Are there others like us who look outward to other worlds and wonder if any are inhabited? In particular, are there other worlds upon which technological civilizations have appeared? Are these civilizations somewhat like ours, or are there vastly superior societies in the stars that are as far advanced over ours as we are over the first primitive multicellular animals that appeared in the seas? Or are we alone?

Within the last couple of decades, human beings have acquired the ability to make their presence known to intelligent species elsewhere in the cosmos. At the present time, four unmanned spacecraft are headed out of the solar system toward the stars. Although the prospect of their ever being found is remote, they carry messages intended for extraterrestrial intelligences. In 1974, a powerful radio transmitter was used to beam a microwave signal to a distant star cluster at the edge of the galaxy. It carried a short message describing our form of life to anyone out there who might be listening. With the advent of advanced electronic technology, it is possible to send a radio

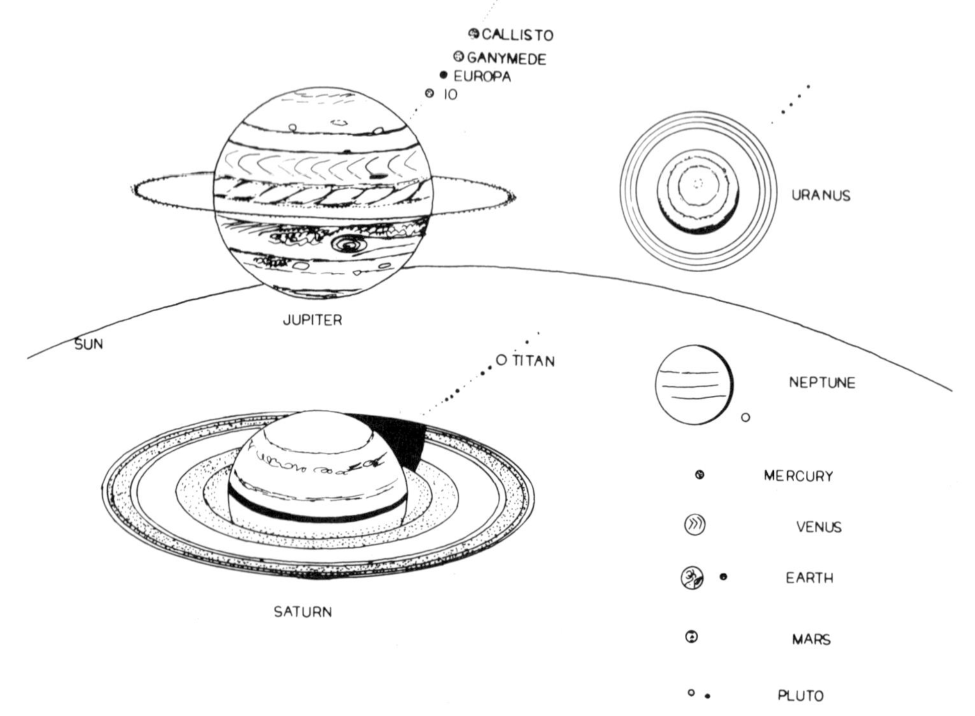

FIGURE C. Relative sizes of the planets. Part of the solar disk is shown in the background.

message to another civilization located anywhere in the galaxy that has a technology at least as advanced as that currently found on Earth. Humanity may be at the threshold of one of the pivotal points in its history: the establishment of contact with alien intelligent creatures.

But we look outward to the stars and see no obvious evidence of extraterrestrial intelligence. Where are they? The existence or nonexistence of alien technological societies is critically relevant to our own destiny upon this Earth. We humans have inherited dangerous evolutionary traits from our reptilian and mammalian ancestors that place our survival in grave peril once we have access to highly advanced technology. Perhaps there are many worlds in the universe which evolved advanced societies that rose to brilliant levels of achievement only to destroy themselves in a few short years through their misuse of technology. If such is the fate of all technological societies, humanity stands very little chance of survival. On the other hand, perhaps there are some technological civilizations that have managed to survive their own crises of technological adolescence and have gone on to build advanced and stable societies lasting for millions or even billions of years. There likely is much that they could teach us about the art of survival. If someone else in the stars has managed to endure, perhaps we also have a chance.

ACKNOWLEDGMENTS

The idea for this book came from a course in Interstellar Communication that I taught when I was a member of the faculty at the Illinois Institute of Technology. The course was originally suggested to me by Professor Leonard Grossweiner, chairman of the Physics Department. I benefited greatly from interaction with the students during the teaching of this course, and many of the ideas developed during this time appear in this book.

I have benefited greatly from parents who encouraged my curiosity and from teachers in high school and college who stimulated my growth. I particularly wish to acknowledge Karen Davé. Her encouragement and criticism have played a crucial role in helping me to bring this work to fruition.

Thanks go to Rosemarie Antoon, who shepherded the preliminary manuscript of this work through the process of evaluation and eventual acceptance by Prentice-Hall. Mary Kennan, my editor, gave several crucial suggestions that were of invaluable assistance in editing the manuscript down to a workable size.

Thanks also to Chris McMorrow, Jean Karash, and their colleagues for the close editing and checking of the manuscript for accuracy and internal consistency. Any errors of omission or commission are, of course, mine.

CHAPTER ONE

LIFE IN THE SOLAR SYSTEM

The space probe had cruised for thousands of years through the cold reaches of interstellar space. Its mechanical intelligence lay dormant, awaiting a stirring from its long electronic sleep. The wise and ancient civilization that had built the craft had been lonely; its citizens had hungered for contact and interaction with other intelligent living creatures like themselves. Many millions of years ago, they constructed this superintelligent robot probe and dispatched it to the stars in a journey of exploration to seek out new civilizations. By now, it had visited many solar systems. It found most planets to be barren and desolate. Life could never be a part of their destinies. A few planets had life forms, but none had yet evolved beyond the most primitive level. No planet had yet been found with an intelligent civilization capable of high technology. The machine intelligence of the probe never tired; the search pressed onward.

The probe now found itself approaching a new target, a single yellow star at the edge of a galactic arm. The warmth of the star brought the ship's electronic brain to life once again. Would this system have the long-sought intelligent creatures, or would it be yet another disappointment?

The ship's sensitive instruments probed the approaching star and found that it had planets. It altered course and passed near the outermost world. It was a frigid world made of frozen methane, with a large moon orbiting nearby. No life was possible here. A larger planet approached, a rapidly spinning greenish world with two moons. Sensors probed the planet and found only methane, ammonia, and hydrogen gas under its clouds. Temperatures and pressures in the interior were so high that no life could survive. The probe pressed onward. A world similar to the previous one loomed ahead, a greenish, hydrogen-rich planet with five small moons. It too had a dense, crushing atmosphere of poisonous gases. It differed by having a darkly colored set of rings. This world too was barren of life. The probe proceeded onward to the inner parts of the planetary system of this new yellow-colored star.

An exquisite ringed world approached. Even though the craft had seen many thousands of worlds, the beauty of this planet stirred its machine consciousness. This was a giant gaseous planet, composed mainly of hydrogen and helium. As the craft approached, richly colored bands could be seen in the upper atmosphere. The atmosphere of hydrogen was so dense and hot that no life could survive. A beautiful world, but sterile.

The probe passed through the ring plane, then spotted a single large, orange-colored moon. It had a dense atmosphere; perhaps it was warm enough at the surface for life to survive. In eager anticipation, the craft probed the surface. Alas, it was a frozen wasteland forever hostile to life. The probe pressed even closer to the rapidly warming yellow sun.

Another giant gas world approached, one even larger than the exquisite ringed planet just visited. Like that of so many other gas giants that had been encountered, the atmosphere was hostile to life. No intelligent technological society would ever come to this giant world. The sensors found four large moons orbiting this planet; perhaps they were better prospects for life. However, none had atmospheres. The outer two were ice-covered worlds scarred with the remains of giant impact craters. The second moon was an ice-covered world crisscrossed with cracks. The innermost moon was a brightly orange-colored sphere vigorous with volcanic activity. But no life anywhere.

The innermost regions of this planetary system lay ahead. Life might have a better chance there. A small, reddish-colored world with two tiny moons approached. It had a thin atmosphere; perhaps some regions on its surface might be able to support life and intelligence. The probe pressed closer. The surface was scarred with giant impact craters and large extinct volcanoes. This world was once covered with liquid water, but the planet was so far from its sun that all its water had long ago frozen solid. If this world ever had life, it must have been extinguished many years ago.

The probe approached closer and closer to the yellow sun. Another world approached, a planet completely covered with dense clouds. It was a promising prospect; it was about the same size as the Builder's home planet and had a dense atmosphere. As the probe neared this perpetually clouded world, the sensors found the surface to be an inferno of superheated steam. Tragically, this promising planet was too close to its sun. Life, if it had ever been present, perished many years earlier.

Yet another world approached. This was a slowly spinning, airless planet with a densely cratered surface. It was so close to the yellow star that the temperature at the surface must be near that of molten lead. No life could ever survive here.

Like so many others, the system of this yellow star had turned out to be lifeless and barren. In bitter disappointment, the probe prepared to shut down for yet another long cruise through interstellar space to another star. But suddenly its sensors spotted another planet, one that had been missed during the initial approach. As this new world loomed larger, it turned out to be a greenish-colored sphere with a single large moon.

This greenish world, third from the sun, had a warm surface.

There were large oceans of liquid water. The atmosphere was rich in oxygen, the unmistakable signature of life! In eager anticipation, the probe pressed closer. Were intelligence and high technology here as well? Was its lonely search at last over?

The sensors looked down on the surface and saw clear evidence of intelligence: vast sophisticated engineering works. The search was over; the probe had found its long-sought civilization. Would these creatures be eager for interstellar contact, or would they be hostile? The probe's sensitive radio receivers listened for signals transmitted by these beings. Only silence answered.

A growing sense of apprehension gripped the ship. Something was terribly wrong. Why no radio signals? The probe began to enter the thin upper reaches of the atmosphere. And then it found the reason: radioactive dust in the atmosphere all over the world, enough to kill any complex life forms. A surface blackened by the scars of mighty explosions came into view. A full nuclear exchange! In an unspeakable act of madness, this society had destroyed itself.

The probe was too late in coming to this world. Perhaps these creatures could have taught much of value to the rest of the cosmos. But they had squandered 4 billion years of painful evolutionary progress in a few hours of senseless destruction. The probe turned away in despair from this now-desolate world. Perhaps life and civilization await on the next star. The craft prepared to leave this now-dead solar system. The search must continue.

This was a dream, a fantasy about the first contact between Earth and an extraterrestrial civilization. One hopes that it was only a dream, and not a portent of our future. What are the prospects for contact between humanity and extraterrestrial civilizations? We know for certain of no life in the universe other than that which is present on our own planet. Any estimate of the probability of life elsewhere must rely on a good deal of speculation, hopefully intermixed with some sound scientific fact. We humans have recently begun a detailed exploration of the planets and moons in our own solar system. We now know a great deal about the conditions on the surfaces of most of the worlds in our solar system, and some informed answers can now be given about the prospects for life on these worlds. Armed with this information, we can place reliable requirements on the conditions that must be present on a planetary surface to make it fit for life.

The factors that will affect the probability of the existence of life and intelligence elsewhere in the universe will be closely related to those which have governed the appearance of life on Earth as well as to those which have subsequently influenced its evolution. We begin with a description of the Earth viewed within the context of recent planetary discoveries, with special emphasis on the unique properties of our planet which have made it suitable for habitation.

THE EARTH

It can perhaps be said that the most asthetically pleasing products of the space age are the color photographs of the Earth which have been taken from outer space. In these pictures the Earth appears as a bluish-colored sphere interspersed with brownish areas, hauntingly beautiful up against a backdrop of the inky blackness of outer space. Apart from its beauty, there is not much to distinguish the Earth from the other planets in the solar system. Some of the planets are much larger, and some are a good deal smaller. What makes the Earth truly unique is that it is probably the only planet in the solar system that has life.

The Earth's orbit around the Sun is approximately circular, with an average distance from the Sun of 149 million kilometers or 93 million miles. This distance defines the *astronomical unit* (or AU), a scale often used in discussing distances within the solar system. The Earth also rotates about its axis once a day as it revolves around the Sun, bringing alternate light and dark to most points on the globe. The seasons of the year are caused by the fact that Earth's equator is not parallel to the ecliptic (the plane made by the Earth's orbit around the Sun). They are approximately 23.5 degrees apart. As the Earth orbits the Sun, the rotational axis of the Earth always points toward the same spot in the sky (toward the star Polaris in the

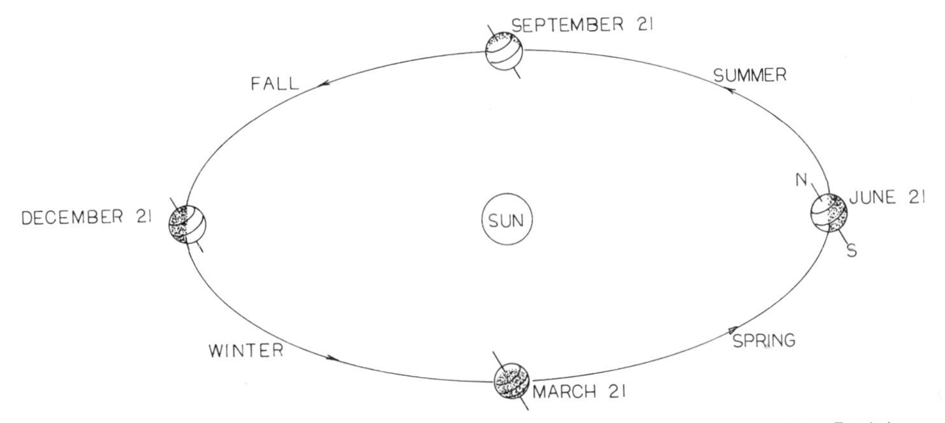

FIGURE 1.1. The seasons of the year are caused by the fact that the equator of the Earth is not parallel to its orbit around the Sun. Since the rotational axis always points to the same direction in the sky, the apparent height of the Sun in the sky will vary throughout the year.

constellation Ursa Minor). As a result the maximum apparent height of the Sun in the sky at any one point on the Earth varies throughout the year. The higher the average height of the Sun in the sky, the larger the average amount of light energy that falls on a unit area of the surface and the higher the average temperature.

The Earth's Interior

The Earth is an approximately spherical body with a mean radius of 6378 kilometers. The average density is 5.5 grams per cubic centimeter (gm/cc). The mean density of the surface material is only about 3.0 gm/cc, so the interior of the Earth must be significantly denser. There are three distinct regions: a *crust,* a *mantle,* and a *core.* The outer crust is about 30 kilometers thick and has a density ranging from 2.7 to 3.0 gm/cc. It is rich in silicate minerals, with aluminum, iron, calcium, sodium, potassium, and magnesium metals being abundant. Below the crust lies the mantle. The very uppermost layers of the mantle just beneath the crust are at a temperature just below the melting point and are able to flow (or "creep") under the application of stress. Some regions of the upper mantle are actually molten, producing lava flows when this material is forced to the surface. The mantle increases in density with depth, reaching a value as high as 6.0 gm/cc at a depth of 3000 kilometers. At that depth, the pressure is 1.5 million atmospheres and the temperature is 2500

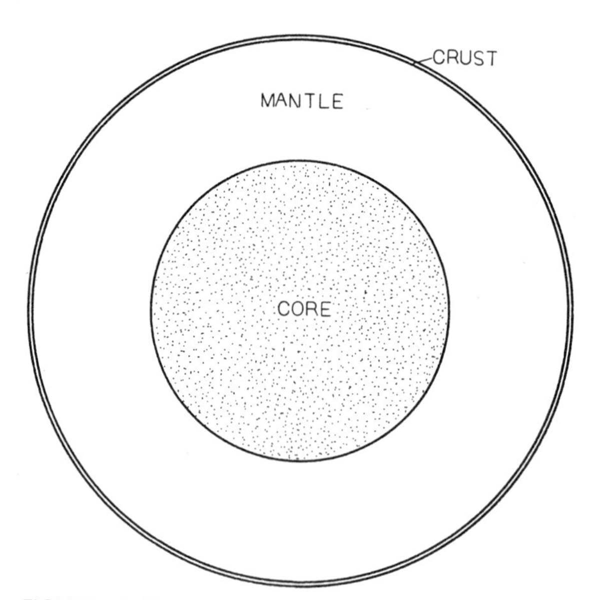

FIGURE 1.2. The interior of the Earth.

degrees Celsius. At depths below 3000 kilometers, the material is hot enough to be entirely in the liquid phase. This is the core. The boundary of the core is quite sharp, and the density at the interface abruptly jumps from 6.0 gm/cc to 9.9 gm/cc. The only known liquid substance with a density this high is molten metal. The core is probably iron, perhaps diluted with four percent nickel and 8-10 percent sulfur.

The Earth possesses an intrinsic magnetic field, of maximum intensity approximately 0.5 gauss at the surface. This field can be visualized as being produced by a giant bar magnet located near the center of the Earth. It is probably caused by persistent electric currents existing deep within the molten core and driven by the combined effects of heat convection and planetary rotation. The Sun emits a steady stream of charged particles (primarily protons and electrons) known as the *solar wind.* The terrestrial magnetic field deflects most of the solar wind particles impinging upon the Earth, preventing them from striking the upper atmosphere. There is a dense region of particle radiation just above the Earth's atmosphere that is confined by the terrestrial magnetic field. Most of this radiation lies roughly in the plane of the Earth's magnetic equator. There are two toriodal-shaped regions where the radiation is particularly intense, one at an altitude of 2000 to 5000 kilometers and the other at an altitude of 13,000 to 19,000 kilometers. These regions are called *Van Allen belts,* after James A. Van Allen, who played a leading role in their discovery and interpretation at the dawn of the space age in the late 1950s. The peak radiation intensity in the outer Van Allen belt is sufficiently high to be a health hazard to astronauts flying through this region. It is therefore a good idea to pass through the outer parts of the Van Allen belts as quickly as possible.

The Earth's Surface

The outermost crust of the Earth is laterally subdivided into two distinct regions, the lower-lying oceanic floors and the higher continental land masses. The oceanic floors are primarily dark-colored silicate minerals rich in calcium, aluminum, magnesium, and iron. The average density of this material is about 2.9 gm/cc. Oceanic crust is rather thin, with an average thickness of only 6 kilometers. The continental land masses are somewhat thicker and are slightly less dense (about 2.7 gm/cc). Continents have lots of coarse-grained, light-colored rocks of volcanic origin that are rich in quartz ($Si0_2$) and potassium aluminosilicate minerals.

One of the major discoveries of this century is the fact that much of the Earth's surface is undergoing constant alteration, destruction,

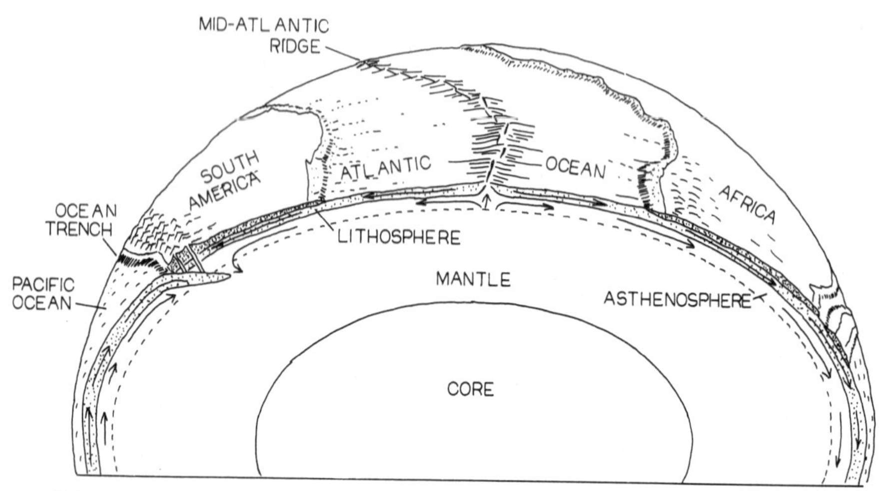

FIGURE 1.3. A slice through the Earth's crust. Africa and South America are being forced apart by rising convection currents along the mid-Atlantic ridge. On the western edge of South America, continental plates are undergoing subduction, giving rise to volcanic activity. From Wyllie (1975). Copyright © 1975 by *Scientific American, Inc.* All rights reserved.

and reformation. The crust and mantle down to a depth of 100 kilometers are divided laterally into ten major continental plates that cover the entire Earth. Plate boundaries are delineated by the long and narrow regions where earthquakes are frequent. The plates "float" on top of a 150-kilometer-thick layer of partially molten mantle. The plates migrate laterally over the surface at a rate of a few centimeters per year, so that the map of the Earth changes significantly over a time period as short as half a million years. Molten lava flows upward through cracks in the surface, forcing adjacent plates apart. As the lava cools, new crust is created. In other regions plates are forced together, forming new mountains. In other regions, one continental land mass actually slides up over another as they collide, with the lower plate being pushed down into the mantle and remelted. There is a continuous exchange of rocky material between the solid outer crust and the molten upper mantle upon which it floats.

Water and Air

Perhaps the most important property of the Earth as far as life is concerned is the presence of large amounts of liquid water flowing freely on the surface. At any one time, about 10 percent of the Earth's water is frozen into solid ice. About 7 percent of the Earth's surface is covered with ice, mostly at the northern and southern poles, with the southern polar cap containing about twice as much ice as the northern.

FIGURE 1.4. The positions of the continents 200 million years ago, as compared to their positions today. From Hallam (1975).

Earth's dense atmosphere is almost as important as water itself in the maintenance of life on this planet. Seventy-eight percent of the atmosphere (by volume) is nitrogen (N_2), 21 percent is oxygen (O_2), and about 1 percent is argon (Ar). There is also a lot of water vapor (H_2O) in the air, the exact amount of which varies with surface location as well as with the season of the year (ranging from virtually zero to as high as four percent). There is a constant exchange of water between the oceans and the atmosphere via the cycle of evaporation, condensation, and precipitation. Several other gases are also present in the atmosphere, but in much smaller amounts. There are about 330 parts per million (ppm) of carbon dioxide (CO_2), 70 ppm helium (He),

18 ppm neon (Ne), 1.5 ppm methane (CH_4), and 1 ppm krypton (Kr). Gases such as hydrogen (H_2), nitrous oxide (N_2O), carbon monoxide (CO), ammonia (NH_3), nitrogen dioxide (NO_2), sulfur dioxide (SO_2), hydrogen sulfide (H_2S), and ozone (O_3) are also found, but only in trace amounts of 0.5 ppm or less.

The air pressure at sea level is defined as one atmosphere. As one climbs above sea level, the pressure steadily decreases. For every 5.5 kilometers increase of height, the pressure drops by one half. The atmosphere is vertically subdivided into five distinct layers, termed *troposphere, stratosphere, mesosphere, ionosphere,* and *thermosphere.*

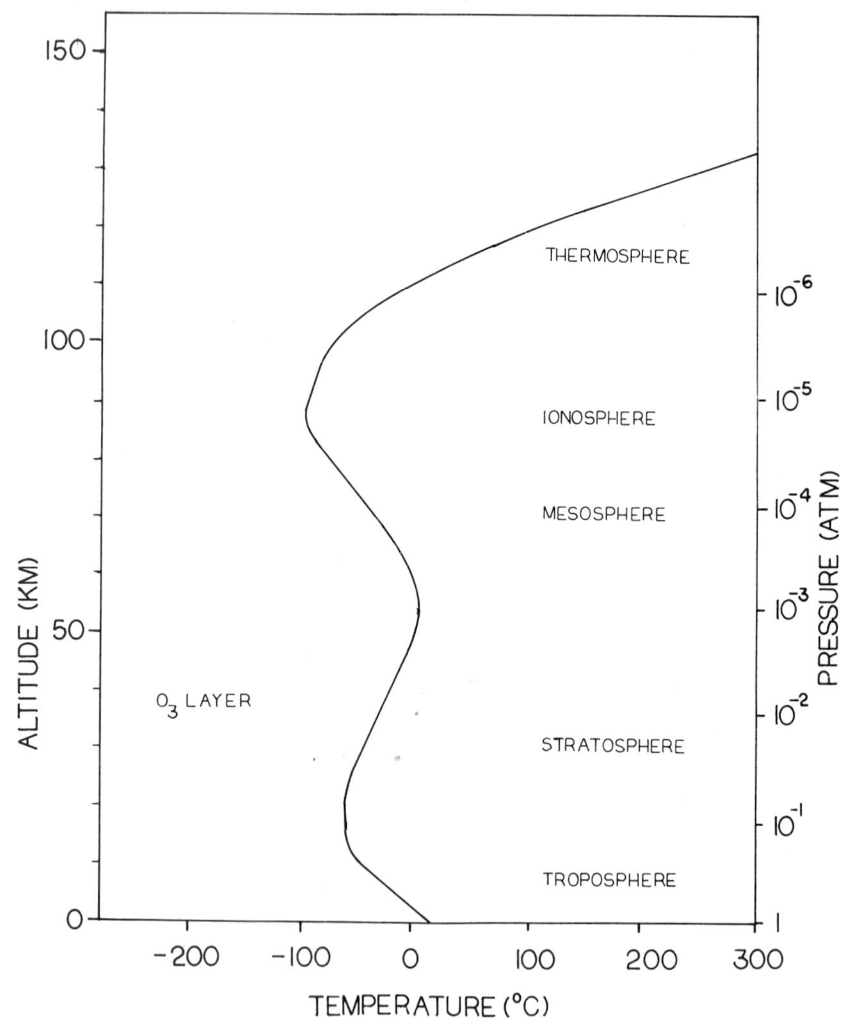

FIGURE 1.5. The vertical profile of the Earth's atmosphere.

The lowest ten kilometers of the atmosphere comprise the troposphere, where all living things reside. The temperature at the surface averages 13 degrees Celsius (55 degrees Fahrenheit). The high temperature of the lower troposphere is a result of the absorption of incident solar energy at the Earth's surface. The average temperature of the troposphere decreases as one gets farther from the surface. At a height of ten kilometers, the temperature is down to −50 degrees Celsius. At this point the stratosphere begins. Above this level the temperature begins to rise with ascent, reaching a value nearly that of sea level at an altitude of 50 kilometers. The energy source responsible for the high temperature of the stratosphere is *oxygen photochemistry*. Low-wavelength ultraviolet light from the Sun breaks up stratospheric oxygen molecules into their constituent atoms. Some of these oxygen atoms combine with other oxygen molecules to form ozone (O_3). This ozone layer absorbs solar ultraviolet light of still longer wavelength. The net effect of oxygen photochemistry in the stratosphere is that ultraviolet light of wavelengths shorter than 3000 angstroms is prevented from reaching the surface of the Earth, where it would be a hazard to life. At a height of 50 kilometers, the temperature begins to decline with height once again, and the mesosphere begins. At 80 kilometers there is a highly variable layer of charged particles hovering in the atmosphere. This is the ionosphere, which is sustained by the absorption of highly energetic X-rays emitted by the Sun. Above the ionosphere, the air temperature begins to rise yet again. This is the thermosphere, which is heated by the influx of highly energetic charged particles coming from the Sun and from cosmic space. By the time an altitude of 150 kilometers is reached, the atmosphere has merged into the near-vacuum of outer space.

The Cycle of Life

Unlike those of the other planets in the Solar System, the Earth's protective air blanket has largely been formed and shaped by biological activity. The best-known of these interactions is the *oxygen–carbon dioxide cycle*. Photosynthetic organisms take some of the carbon dioxide in the air and combine it with water in the presence of light to create sugar for use as food. Oxygen is released to the air as a by-product of this process. Other living beings use this oxygen for respiration, releasing carbon dioxide back into the air for another round of the cycle.

The Earth's oceans and crust also act as important "sinks" for carbon dioxide. Carbon dioxide is highly soluble in water, where it

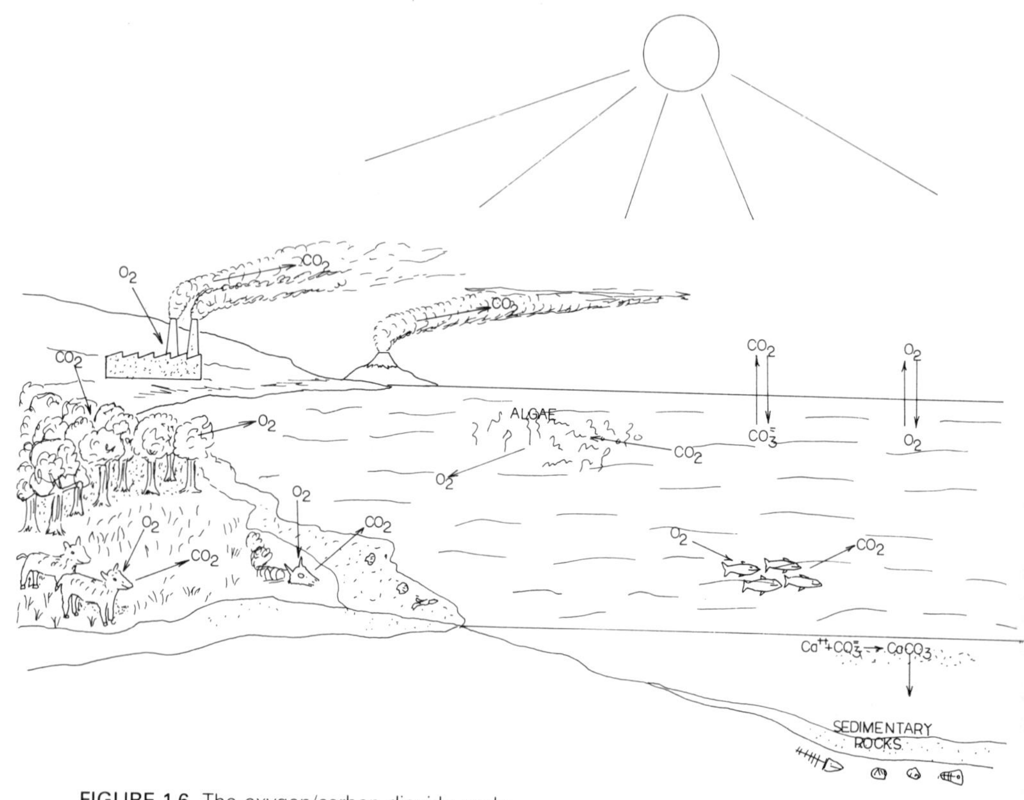

FIGURE 1.6. The oxygen/carbon dioxide cycle.

forms carbonate or bicarbonate ions. These carbonate ions can react with any dissolved calcium ions to form the insoluble compound calcium carbonate, the prime constituent of limestone. This limestone settles to the bottom of the oceans and forms into sedimentary rocks. Over the years, these carbonate minerals have become an important component of the Earth's crust. If all of the carbon dioxide currently "locked" into the rocks were to enter the atmosphere, the pressure would rise to 25 atmospheres.

The Edge of Disaster

The average temperature at Earth's surface is warm enough to keep most of Earth's water from freezing into solid ice but not so hot that the oceans all vaporize. However, the "effective temperature" of the Earth (as determined by the energy balance between the sunlight that is absorbed by the planet and that which is reflected back into space) is only −20 degrees Celsius, far below the freezing point of water. The surface of the Earth is at least 30 degrees Celsius warmer than it should be if it were heated simply by absorbing the solar energy incident upon it. What is responsible for this excess warmth? The cause appears to be a "greenhouse effect" produced by the slight amount of carbon dioxide in the atmosphere. Carbon dioxide is transparent to visible light, but it absorbs infrared light very strongly. When sunlight strikes the surface, part of the solar energy is reflected back into space, whereas the rest is absorbed. The warm surface reradiates this heat energy back into space, but this radiation takes place primarily in the infrared rather than the visible region of the spectrum. The infrared radiation coming from the surface is absorbed by the carbon dioxide in the atmosphere, and the heat energy becomes trapped near the ground. If all of the carbon dioxide were suddenly to vanish from the atmosphere, the average temperature of the Earth would soon cool below the freezing point of water. The lower temperature would cause more of Earth's water to freeze into solid ice, which would increase the overall reflectivity (or albedo) of the planet. The increased albedo would cause a larger fraction of the incident sunlight to reflect back into space, causing the overall termperature to decline still further and even more water to freeze. Eventually, all the water on Earth would freeze. Such a process is called *runaway glaciation.* If this ever happens, all living things on the planet will be killed. A significant increase in the atmospheric carbon dioxide content would be just as unfortunate. If more carbon dioxide were to enter the atmosphere, the temperature would become warmer. Warmer temperatures would force some of the carbon dioxide dissolved in the oceans back out into the atmosphere, trapping still more heat. In the worst imaginable case, the temperature could rise so high that all of the oceans vaporize and all the carbon dioxide is forced out of the rocks, producing a dense, crushing atmosphere of superheated water vapor and carbon dioxide. This rapid uncontrolled temperature rise has come to be known as a *runaway greenhouse effect.* This too would be a disaster, completely destroying all forms of life. The survival of life on Earth has been made possible by the maintenance of a complex equilibrium between the carbon dioxide stored in the atmosphere and that dissolved in the oceans, bound inside living organisms, or locked inside carbonate crustal minerals. If this equilibrium were to be disrupted, catastrophe would result.

The Ages of Rocks

How long has the Earth been here? The ages of the rocks on the surface can be measured by the technique of *radioactive dating.* At the time of the formation of a solid rock from a cooling liquid melt, radioactive elements such as rubidium-87, thorium-232, uranium-238, uranium-235, potassium-40, samarium-147, and rhenium-187 become trapped in the rock. These elements decay at known rates to produce stable "daughter" elements that remain in the rock. By measuring the relative amounts of parent and daughter isotopes, the amount of time elapsed since the initial solidification of a rock can be determined. Most rocks are one to two billion years old, but some are as young as a few million years. The oldest known rocks are in the Isua formation of Greenland. These consist of greenstone rocks and associated sediments dated at 3.8 billion years. In 1983 a few grains of zircon minerals that are 4.1 to 4.2 billion years old were found in Western Australia. Rocks at the bottom of oceanic basins are typically much younger than those in continental land masses, with no deep oceanic crust being any more than 200 million years old. The widely varying ages of earth rocks suggest a complex history of continual melting and reforming of the crust, a process which continues to the present day.

The age of the Earth itself can be determined by the so-called "lead-lead chronometer." Lead in terrestrial rocks exists in four isotopes, Pb-206, Pb-207, Pb-208, and Pb-204. The first three of these are "radiogenic"; they came from the radioactive decay of heavier elements. Pb-204 is "primordial" and was not created by the radioactive decay of any other element. Any primordial lead found in terrestrial rocks must have been present on this planet as long as the Earth itself had existed. Every time a rock melted and reformed, its primordial lead was progressively diluted with more and more radiogenic lead. By measuring the relative amounts of lead isotopes present in minerals, the age of the Earth can be determined. The results seem to indicate that the Earth first formed as a solid, coherent body about 4.7 billion years ago.

AIR, WATER, EARTH, AND LIFE: AN ETERNAL SYMBIOSIS

The Earth and the other planets initially formed from a gaseous and dusty disk of material which accumulated around the Sun nearly five billion years ago. The young Earth probably had five times as many radioactive elements as it does at present, and the radioactive decay caused the interior of the planet to become very hot. The entire Earth must have melted, and any geological record of this early time was irretrievably lost. The denser metallic material sank to the interior to form the core, leaving the lighter silicate material "floating" on top to form the mantle and crust. At this time the entire surface of the Earth must have been covered by an "ocean" of molten lava.

As the rate of radioactive heating in the interior abated, the Earth began to cool. About 4.2 to 4.3 billion years ago, the surface temperature was low enough for a solid crust to form. The interior still remained molten, and intense volcanic activity must have been prevalent almost everywhere. The first crust was probably broken up into many thousands of tiny plates that solidified, remelted, then reformed once again many, many times. Little record of this early period has survived. At this time, the solar system still had a large number of asteriod-sized bodies that had not yet coalesced into planets. They rained down on the surfaces of the planets and moons in the new solar system, excavating large basins and impact craters. The young Earth probably looked a lot like the Moon does now, with a surface saturated with densely overlapping craters. However, the intense geological activity on Earth's surface soon erased most of these craters, leaving behind only the barest hint of their existence. The most intense period of meteoric bombardment ended about 4 billion years ago, and only an occasional small meteorite strikes the Earth today.

The First Air

The Earth probably acquired some sort of atmosphere immediately after it first formed as a solid body. This atmosphere was gathered up from the gases in the solar nebula. It must have been largely hydrogen and helium, since these gases are found today in abundant amounts in the atmospheres of the Jovian planets. Jupiter was large enough to retain its primal atmosphere, but the Earth was not, and the terrestrial hydrogen-helium proto-atmosphere was quickly lost to outer space. Earth probably did not begin to acquire a permanent atmosphere until somewhat later. When the interior began to melt, some gases were baked out of the molten rocks and forced to the surface. The exact chemical composition of this early atmosphere is still a matter of hot controversy. The first systematic model of Earth's primeval atmosphere was proposed by Harold Urey of the University of Chicago during the early 1950s. He suggested that Earth's first permanent atmosphere had to have reflected the composition of the original solar nebula. The simplest gaseous compounds which can be synthesized from an atomic mixture of solar composition and still be retained for any appreciable length of time by Earth's gravitational field are water, ammonia, methane, and hydrogen sulfide. There was no free molecular oxygen. This mixture of gases is classified by chemists as being highly reducing. Urey's model of the primitive atmosphere was accepted by virtually all scientists for many years as being a reasonable working hypothesis. Recently, however, a somewhat different model has gained many adherents. This model proposes that the primordial atmosphere was produced from the vapors given off by internal geochemical processes driven by the intense heating. These gases must have been chlorine, water, carbon dioxide, nitrogen, and hydrogen sulfide, since they are emitted in copious amounts during present-day volcanic eruptions. This mixture of gases is considerably less reducing than the ammonia-methane mixture proposed by Urey.

There are carbonate minerals in the oldest known Earth rocks, suggesting the existence of a carbon dioxide atmosphere at the time of their formation. Even if Earth's atmosphere had originally possessed a large amount of ammonia and methane, these gases probably could not have been retained for very long. The ammonia in the atmosphere must have been particularly short-lived. Ammonia is extremely soluble in water, and most of it must have very quickly "washed out" of the atmosphere at the time the first oceans formed. What little ammonia remained was *photodissociated* by solar ultraviolet light. Ammonia molecules were disrupted into their component hydrogen atoms and nitrogen radicals. The hydrogen escaped into space, but the nitrogen radicals remained behind and combined with each other to produce molecular nitrogen (N_2). The photodecomposition of the primordial ammonia may be the major source of the nitrogen gas found in today's atmosphere. *Water ultraviolet photolysis* also played a critical role. The early atmosphere must have been much more humid than today's air because of the high temperature, and many water molecules were torn apart by the ultraviolet light to produce oxygen and hydrogen radical species. The hydrogen escaped into space, but the oxygen radicals remained behind to attack the methane molecules, creating carbon dioxide. By means of such complex photochemical processes, any atmosphere of methane and ammonia would have been quickly converted by ultraviolet light into carbon dioxide and nitrogen. Any primitive terrestrial atmosphere of methane or ammonia must have been extremely short-lived, if it even ever existed at all.

Eventually, the temperature at the surface dropped below 100 degrees Celsius as the Earth continued to cool. At this time, the water vapor in the air began to condense to form liquid water. It must have continuously rained all over the world for many millions of years. This sudden flood of water rapidly accumulated in the lower regions on the surface, forming the first oceans. Extensive oceans must have been present at least 3.8 billion years ago, as some of the oldest rocks yet discovered show evidence of large amounts of water having been present during their formation. Liquid water has been present on Earth in significant amounts ever since.

It is now known that the Sun has been slowly but steadily increasing in luminosity for the past four billion years. Various estimates predict that the Sun of 3.5 to 4.0 billion years ago was 15 to 30 percent dimmer than it is now. This introduces a troubling paradox: why then is there clear geological evidence for the presence of water in the liquid state at this early time? The effective temperature of the Earth should have been 8 percent lower 4 billion years ago than it is today, sufficiently low to have kept all water frozen solid for the first 2 billion years of Earth's history. This obviously did not happen. In point of fact, much geological evidence seems to indicate that the average surface temperature at this time could have been as high as 50 to 70 degrees Celsius. What was responsible for this odd state of affairs? The cause is thought to have been Earth's ancient reducing atmosphere. The ammonia, methane, water vapor, and carbon dioxide presumed to have been in the primeval atmosphere were able to trap large amounts of heat via the "greenhouse effect." The ammonia and methane have gradually been removed from the atmosphere and the carbon dioxide has been incorporated into the limestone crust and into living beings, resulting in a gradual cooling trend over the past 4 billion years. This trend may continue in the future, provided humans do nothing to disturb it.

The Advent of Life

The subsequent geological history of the Earth cannot be separated from the evolution of life on this planet. The earliest period of Earth's geological history is called the *Archean;* it is said to begin with the formation of the Earth's first crust and end with the acquisition of the first permanent continental plates approximately 2.5 billion years ago. However, because of the intense volcanic activity of that era, very few Archean geological formations have survived.

For a long time it was thought that there was no life present on Earth until the time of the Cambrian era, about 570 million years ago. Rocks laid down at this time show abundant fossil evidence of the presence of a rich variety of animal and plant life. Earlier rock strata seemed to have no fossils at all, leading people to conclude that our planet had been completely lifeless before that time. Today it is known that earlier geological strata do indeed contain fossils, but they are microscopic in size. Microfossils as old as 3.5 billion years have been found in Archean deposits. For three billion years, it appears, the only life present on Earth was of the single-celled variety. This entire period of single-celled life is referred to as the *Cryptozoic* eon (the eon of "hidden life") or simply as the *Precambrian.* It includes both the Archean era and the Proterozoic era (the era of "primitive life").

The first living cells must have been very primitive. They were undoubtedly *procaryotic,* lacking distinct cell nuclei. These early creatures were probably little more than aqueous solutions of organic molecules housed inside membranous bags. They were probably entirely *heterotrophic,* drawing all needed energy and building-block molecules from the surrounding water. The energy-conversion process used by these creatures was probably anaerobic fermentation of glucose, a process still used today by some more primitive single-celled organisms that live in oxygen-poor environments.

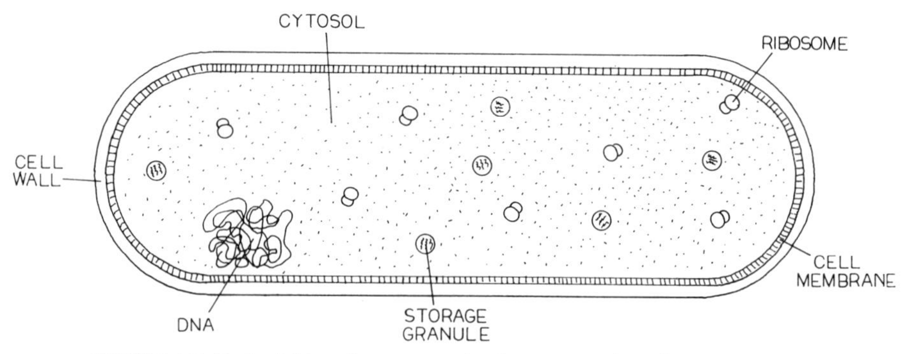

FIGURE 1.7. Escherichia coli, an example of a procaryotic cell.

As the new anaerobic heterotrophs evolved and proliferated, the organic nutrients dissolved in the primordial oceanic "soup" must have gradually become depleted. If this had continued unchecked, the nutrients would have become exhausted and life would have come to an abrupt end. This catastrophe was averted by the evolution of photosynthesis. Organisms appeared that were capable of absorbing visible light from the Sun and using the energy to convert dissolved carbon dioxide into glucose. The photosynthesized glucose ($C_6H_{12}O_6$) was then stored within the bodies of these organisms and was anaerobically degraded whenever food was needed. These early photosynthetic organisms employed a primitive form of photosynthesis, one still used today by green sulfur and purple bacteria that live in anaerobic environments rich in sulfur. These bacteria create glucose out of carbon dioxide and hydrogen sulfide via the reaction:

$$6CO_2 + 12H_2S \xrightarrow{\text{light}} C_6H_{12}O_6 + 6H_2O + 12S \quad (1\text{–}1)$$

The energy required to drive this reaction is provided by the absorption of sunlight.

Oxygen and Evolution

Although there was probably no biological production of oxygen before the mid-Proterozoic (about 2 billion years ago), there was nevertheless a steady generation of molecular oxygen all throughout the Archean via the ultraviolet photolysis of water vapor in the upper atmosphere. The oxygen content was kept rather low by reaction with the reduced gases in the atmosphere. However, some did manage to enter the water. The rising oxygen content in the oceans spelled trouble for the microorganisms living there, as oxygen was highly toxic to them. They had two choices: either move to a more anoxic environment or else develop biochemical apparatus to protect against the toxic effects of oxygen. Evidence for a steadily rising oxygen level throughout the Cryptozoic can be found in present-day procaryotic cells. They have widely varying tolerances for oxygen. Some primitive bacteria are unable to grow in the presence of oxygen and are actually killed by sufficiently high levels of the gas. Others can tolerate oxygen, but will grow much faster in its absence. Still others will grow better with oxygen than without, but only at a much lower concentration than that which is currently present. Finally, there are a few procaryotes that cannot survive without oxygen. We are apparently seeing the results of a slow evolution and adaptation of procaryotic cells to an ever-increasing oxygen concentration.

About two billion years ago another important turn was taken. Procaryotic organisms appeared that employed a more advanced form of photosynthesis, one that could use water rather than hydrogen sulfide as a hydrogen donor:

$$6CO_2 + 6H_2O \xrightarrow{\text{light}} C_6H_{12}O_6 + 6O_2 \qquad (1\text{–}2)$$

These cells were the first *cyanobacteria* and were the prototype for all of the green algae in the oceans as well as for all green plants. The advantages of a photosynthetic process using water as the primary reactant are obvious. Water is certainly far more abundant than hydrogen sulfide ever was, and the cyanobacteria rapidly grew and prospered. Especially important for later developments was the fact that one of the "waste products" emitted by this new photosynthetic process was oxygen gas.

The cyanobacteria were so successful and proliferated so swiftly that the geological record clearly shows evidence of their appearance. The sudden burst of oxygen which was emitted into the oceans reacted with inorganic ferrous ions dissolved in the water, creating insoluble ferric oxide, which settled to the ocean bottoms. This material formed the so-called "banded iron formations" that are the primary sources of iron ore. These formations are all approximately two billion years in age and have been traditionally used to date the first appearance of the cyanobacteria. Virtually all of the reduced iron was cleared from the oceans in less than a hundred million years. Once the bulk of the reduced iron was gone from the oceans, some of the dissolved oxygen generated by the cyanobacteria was forced out into the atmosphere, where it reacted with the final remnants of the reduced gases still present in the atmosphere. The oxidation of the reduced gases in the atmosphere resulted in the removal of an efficient heat-trapping agent that had kept the Earth's temperature warm all throughout the Archean. The geological record shows a significant worldwide cooling trend at this time, and major glaciation appeared for the first time. Once the reduced gases were gone, the atmospheric oxygen concentration rapidly began to rise.

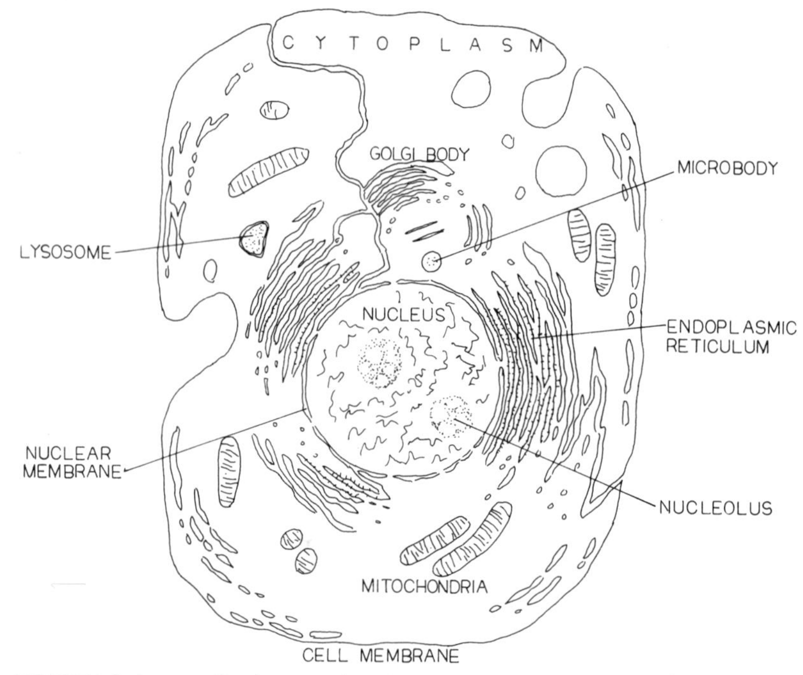

FIGURE 1.8. A generalized eucaryotic cell. From Brachet (1961). Copyright © 1961, *Scientific American*, Inc. All rights reserved.

Approximately a billion and a half years ago, the increased availability of oxygen made possible another watershed in the history of life. This was the appearance of the first *eucaryotic* cell. Eucaryotic cells are generally larger than their procaryotic predecessors and have membrane-surrounded structures known as *organelles* in which the important biochemical reactions are compartmentalized. In particular, the genetic apparatus of the cell is housed within a structure known as the *nucleus.* Without exception, the cells which make up complex multicellular organisms such as ourselves are eucaryotic. Unlike the procaryotes, all contemporary eucaryotic cells (with only a few exceptions) have an absolute requirement for oxygen and will die if kept for too long in an anaerobic environment. Many of the essential molecules used by eucaryotic organisms require the presence of molecular oxygen for their biosynthesis. Eucaryotic cells reproduce by mitosis, which cannot take place in the absence of oxygen. The formation of collagen, a fibrous protein that is the primary structural element of skin and connective tissue in multicellular animals of all types, is dependent on the presence of adequate amounts of oxygen. Without oxygen, the existence of creatures with muscles or brains or any type of multicellular life at all would be impossible.

Perhaps the most important innovation of all was the eucaryotic cell's ability to use molecular oxygen to provide energy. This process is known as *respiration* and involves the use of oxygen to decompose glucose via the reaction:

$$C_6H_{12}O_6 + 6O_2 \longrightarrow 6CO_2 + 6H_2O \qquad (1\text{–}3)$$

This process is the inverse of aerobic photosynthesis (Equation 1–2), since glucose is degraded back to carbon dioxide and water. Respiration yields eighteen times as much energy per glucose molecule consumed as does anaerobic fermentation. By breaking down glucose into water and carbon dioxide, virtually all of the biologically useful energy can be recovered. A cycle is set up involving a coupling of photosynthesis with respiration: Energy from the Sun is used to synthesize glucose from water and carbon dioxide, the energy thereby stored in glucose is liberated inside a living organism by respiration, which in turn releases water and carbon dioxide back to the atmosphere for another round of the cycle.

The Coming of Intelligence

All throughout the Precambrian life consisted exclusively of single-celled organisms. Approximately six hundred million years ago there was a sudden and unprecedented proliferation of a large variety of multicellular (or *metazoan*) organisms. The advent of metazoan life is said to mark the beginning of the *Phanerozoic* eon, which has lasted to the present. The Phanerozoic is divided into three different geological eras, termed *Paleozoic, Mesozoic,* and *Cenozoic.* The first animals to appear were worms and jellyfish, but hard-shelled creatures such as the trilobite soon evolved. Their shells provided a base for strong muscles and helped ward off predators, but the shells also made these creatures clumsy and slow-moving. Approximately 480 million years ago, the first fish appeared. They traded hard protective shells for supple and strong bodies with internal skeletons that gave them enhanced speed and maneuverability in pursuing prey and escaping from danger. The fish were so successful that they rapidly rose to dominance in the oceans, a position which they retain to the present day.

Approximately 400 million years ago, a few hardy bacteria, fungi, and lower plants were able to establish themselves on the land. The warmth and security of the ever-nourishing ocean were traded for the harsh and barren land with its hot and dry summers and freezing winters. This move was made possible only by the existence of an oxygen-rich atmosphere coupled with the presence of an effective ozone shield. The oxygen content in the atmosphere at this time may have been as high as a tenth of the present level. The first animals ashore were the *arthropods,* represented today by such creatures as insects, spiders, scorpions, and crabs. They were soon followed by the *amphibians.* The first amphibian had evolved from certain species of lobe-finned fish that had the ability to gulp air. Such fish could propel themselves along the ground for short distances, moving from one stagnant pool to another. As the years passed, the lobed fins gradually evolved into legs.

The body of the amphibian must be periodically moistened, and its egg must be laid in water. It is a composite creature, at home neither on the land nor in the sea. Three hundred million years ago, creatures appeared that were much better adapted for full-time life on the land. They had tough hides that preserved the water inside their bodies, and they laid eggs with firm shells that preserved an aqueous environment for the embryo inside independent of their being in the water. These creatures were the *reptiles.* The reptiles rapidly grew to dominance and entirely displaced the large amphibians. Reptiles do have one important disadvantage: They are cold-blooded creatures. A reptile's body temperature is very nearly equal to that of the ambient temperature. When it is too warm, it must find a cooler spot or die. When it is too cold, it becomes torpid and incapable of sustained activity.

The Mesozoic era is best known for the appearance of the dinosaurs. They evolved from some of the early reptiles and were one of the most successful of nature's experiments. They dominated the Earth for 150 million years. Some of them were the largest terrestrial creatures ever to live. A few even showed some signs of intelligence. Then suddenly, in the twinkling of an eye, they were all gone. The cause of the extinction of the dinosaurs is believed to be the impact of a gigantic meteorite 65 million years ago which kicked up enough dust into the upper atmosphere to create a perpetual night lasting for a couple of years. The darkness and the cold swept away about half the genera existing on Earth. No terrestrial vertebrate larger than 25 kilograms was able to survive.

The disaster which cleared away the dinosaurs opened up the world to the mammals. Unlike reptiles, mammals are warm-blooded, have body hair, possess a respiratory diaphragm, and give birth to live young. Mammalian young undergo a long period of prematurity during which they are utterly dependent on their parents for food and protection. This forced the development of some sort of family structure to improve the chances for survival. Perhaps the most valuable asset of the early mammals was their enhanced intelligence; their brains were on the average three to ten times larger in proportion to their weights than the brains of dinosaurs.

The era of the mammals, known as the Cenozoic, has lasted to the present day. The removal of the competition provided by the

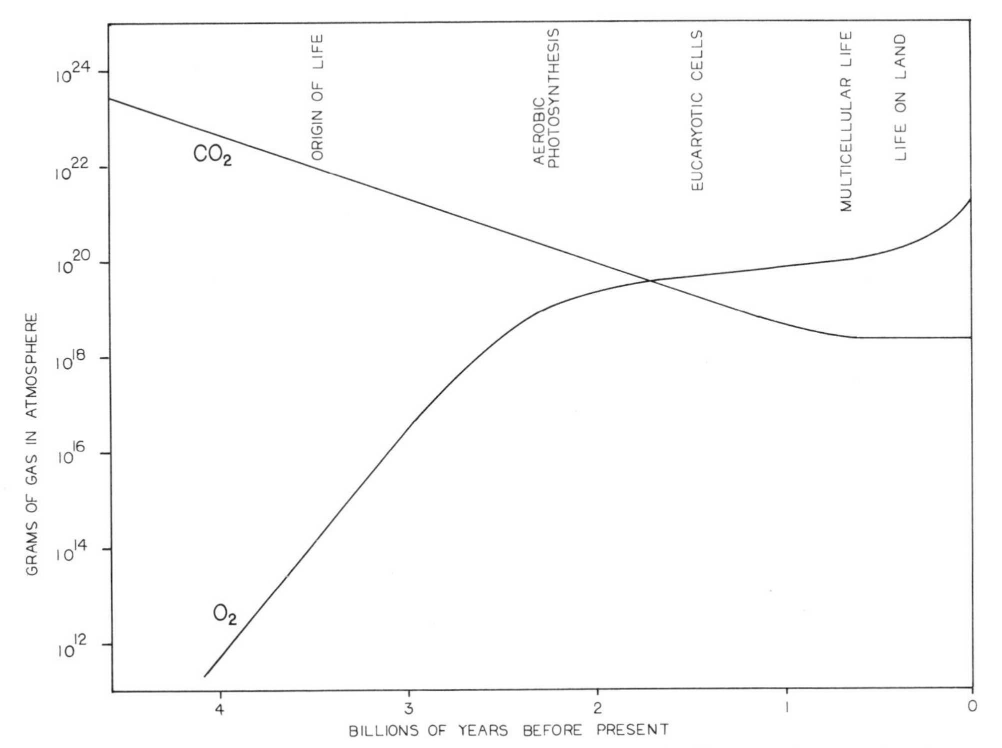

FIGURE 1.9. The oxygen and carbon dioxide content of the Earth's atmosphere as a function of time. It is assumed that all of Earth's carbon dioxide was in the atmosphere at the time of the initial formation of the planet.

dinosaurs allowed thousands of different species of mammals to evolve and expand over the entire Earth. The explosive expansion (or *radiation*) of the mammals is one of the most striking events in all evolutionary history. One of the many branches of this mammalian radiation is especially significant, because it led to us. It began with tree-dwelling lemurs, continued with monkeys, then went to apes, and eventually produced human beings. The supple hands evolved by early humans gave them the manual dexterity which enabled them to make tools. They invented technology. Their brains grew larger, and they developed curiosity about their world. Their descendants built cities and established civilizations. They invented metallurgy and the written word. They established science and mathematics. Human technology grew and expanded and the sons of Man now occupy virtually every possible niche on the planet. Now we are looking outward to the other planets in search of new worlds to conquer.

OTHER WORLDS, ALIEN LANDSCAPES

Moon

Is it possible that the Moon, like the Earth, is inhabited by living creatures? The Moon is an entirely airless world orbiting 384,400 kilometers from Earth. It perpetually turns the same face toward our planet as it rotates, a result of strong tidal forces exerted by the Earth. The Moon is roughly 1.2 percent as massive as Earth, and its diameter is 3480 kilometers. The lunar soil samples brought back to Earth by the Apollo landing teams indicate that the Moon is made up of the same material as the Earth, with silicates of such metals as iron, aluminum, titanium, and magnesium being the most abundant minerals present. The Moon is significantly less dense than Earth, which must mean that it has a much smaller central core. There is no significant magnetic field, and no radiation belts.

The entire surface of the Moon is pockmarked with large craters. They were formed approximately 4 billion years ago by the impacts of large meteorites. Earth also experienced meteorite impacts, but geological activity has erased virtually all of the craters that were formed. Since the Moon has been so much less geologically active, most craters have survived.

The darker areas of the Moon are known as *maria*. They are roughly circular regions made up of solidified lava. The maria were created by the impacts of large asteroids early in the history of the solar system. The impacts were sufficiently severe to excavate large depressions in the lunar surface and penetrate a significant distance into the lunar crust. Approximately 3.5 billion years ago, lava from the lunar mantle flowed upward and filled these impact basins. The solidified lava formed the dark maria seen today. Shortly afterward, all geological activity ceased, leaving the Moon a cold, dead world forever after.

The Apollo lunar landings demonstrated that there is no life whatsoever on the Moon, nor is there evidence of any living beings ever having been present. There is no water on the surface, either in liquid or solid form or in the form of water of hydration locked within surface rocks. The Moon has probably been bone-dry ever since it first formed a solid crust 4.2 billion years ago. The Moon is too small to maintain any permanent atmosphere, and the surface is therefore subjected to extreme temperature variations. At the Apollo 11 landing site in the Sea of Tranquility, the temperature reached a broiling 70 degrees Celsius during the long lunar day. However, the temperature dropped as low as −50 degrees Celsius during the night. The absence of an atmosphere also allows a steady rain of tiny meteorites to reach the surface, slowly but steadily eroding away the surface rocks and turning over the soil. The lack of an atmosphere also means that the surface is exposed to low-wavelength ultraviolet light from the Sun, which would quickly disrupt any complex organic molecules that happen to form. Intense X-rays and gamma rays that are emitted during solar flares and during the active periods of the Sun's cycle of sunspot activity can also reach the surface. The Moon also lacks a significant magnetic field, so the surface is exposed to the full force of the solar wind. This wind has a long-term corrosive effect on any objects exposed to it. Under such severe surface conditions, it is little wonder that living beings have never evolved on the Moon.

Mercury

Mercury is the closest planet to the Sun, a distance of only 0.39 AU away. It has a mass 5.6 percent of that of Earth and a diameter of 4880 kilometers. The density of Mercury is about the same as Earth's. The internal structure of Mercury must then be quite similar, with a dense metallic core rich in iron and nickel and an outer crust and mantle rich in silicate minerals. Superficially, Mercury resembles the Moon. It has a large number of impact craters, indicating that it too underwent a period of intense meteorite bombardment in the distant past. There are large impact basins and smooth intercrater plains. Mercury is a world like the Earth on the inside, but like the Moon on the outside.

The length of the day on Mercury is 58.7 Earth days. The rotational period is suspiciously close to exactly two-thirds of the length of the Mercurian year. As Mercury goes around the Sun twice, it spins on its axis three times. It is extremely unlikely that such a day/year relationship is accidental. Gravitational interactions with the Sun must have locked the planet into its present resonant rotational state very early in the history of the solar system.

As far as life is concerned, much of what has been said about the Moon is equally valid for the planet Mercury. There is no atmosphere and no evidence of any surface water. The surface environment on Mercury is even harsher than that on the Moon; indeed, it is probably the harshest of any in the solar system, with daytime temperatures rising as high as 480 degrees Celsius and nighttime temperatures dropping to −170 degrees Celsius. These sharp temperature extremes are created by Mercury's close proximity to the Sun, coupled with its exceedingly long day. The ultraviolet irradiation on Mercury is ten times as strong as it is on the Moon. Mercury does have a magnetic field, but this field is too weak to prevent an appreciable amount of the solar wind from reaching and damaging the surface. The possibility of even the simplest of organisms being present on Mercury is exceedingly remote.

Venus

Next out from the Sun is the planet Venus, 0.72 AU away. In size, mass, and density, Venus is a virtual twin of the Earth. The planet appears in the telescope as a featureless yellowish disk. The reason for this appearance is the presence of a dense atmosphere which perpetually blocks the surface from view. This atmosphere is largely carbon dioxide gas, intermixed with lesser amounts of such gases as sulfur dioxide, nitrogen, and water vapor. There is a layer of clouds 15 to 45 kilometers above the surface, consisting mainly of droplets of water rich in sulfuric acid. The temperature at the surface is an enormously high 480 degrees Celsius, and the pressure is 90 atmospheres. Venus's surface is hotter than Mercury's dayside hemisphere, even though it is twice as far from the Sun.

The Runaway Oven. The cause of the high surface temperature on Venus is believed to be a "runaway greenhouse effect." During Venus's early history, the planet may have been a lot like Earth, with a cooler temperature and a thinner atmosphere. There may have been lots of liquid water flowing on the surface, enough to have sustained some primitive life forms. The closer proximity of the Sun caused Venus's surface temperature to become so warm that an appreciable amount of liquid water vaporized into the atmosphere. The water vapor trapped solar energy near the surface, causing the temperature to rise

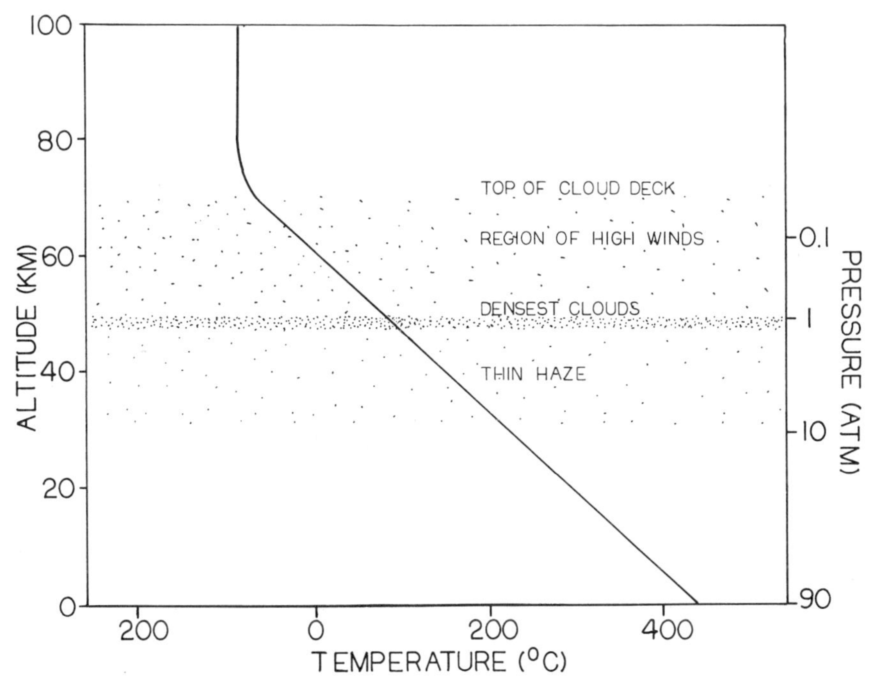

FIGURE 1.10. The vertical profile of the Venusian atmosphere. From Moore and Hunt (1983). Copyright 1983, Mitchell Beazley Publishers, Ltd. Used by permission.

still further. Eventually, the temperature rose so high that an appreciable amount of carbon dioxide was "baked out" of the surface rocks. This outgassing of carbon dioxide trapped still more heat, forcing the temperature to rise still further and driving still more carbon dioxide out of the rocks. The process did not stop until all of the water was vaporized and all of the carbon dioxide was driven out of the rocks. It is interesting to note that the total amount of carbon dioxide currently found in the Venusian atmosphere just about matches the total inventory of terrestrial carbon dioxide. The difference between the two worlds is that on the Earth most of the carbon dioxide is stored within crustal rocks and on Venus it is all out in the atmosphere.

Compared to Earth, Venus is almost completely dry. The planet has less than a thousandth of the amount of water on Earth. Venus must at one time have had nearly as much water as the Earth does now, but lost most of it at the time of the runaway greenhouse disaster. Earth's water has remained largely in the liquid state at the surface, but all of Venus's water was forced into the atmosphere by the high surface temperatures. Once it reached the upper atmosphere, the water vapor was exposed to ultraviolet light from the Sun. This light disrupted water molecules into their constituent oxygen and hydro-

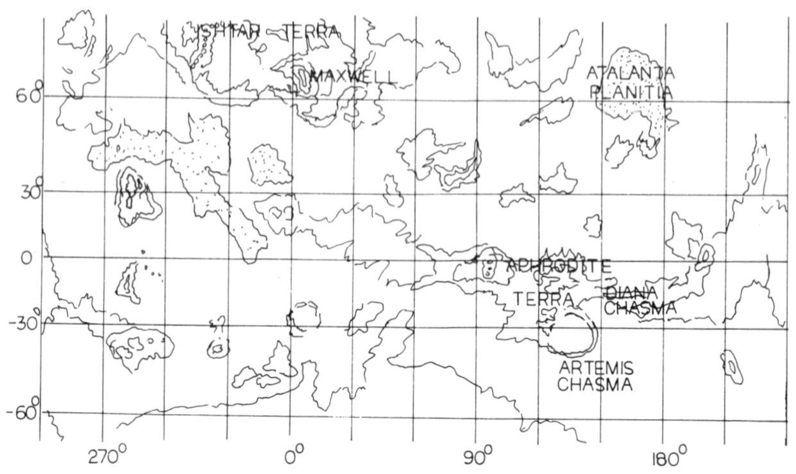

FIGURE 1.11. Radar map of the surface of Venus. Based on map drawn by U. S. Geological Survey.

gen atoms. The hydrogen atoms escaped into space, but the oxygen atoms remained behind to combine with iron atoms in the crust to produce iron oxide.

The Cloaked Surface. Because of the thick clouds, not much is known about the surface morphology of Venus. Earth-based and Venus-orbiting radar installations have been able to map some of the surface features down to a resolution of 10 to 20 kilometers. The radar maps show that Venus is somewhat flatter than the Earth, with at least 60 percent of its surface falling within a height interval of only one kilometer. Three distinct types of terrain have been identified. The first of these is the *lowlands,* comprising approximately 27 percent of the surface. A prominent lowland feature is Atalanta Planitia, a roughly circular depression approximately the size of the Gulf of Mexico. It may be the remains of an ancient impact basin. The second type of terrain on Venus is the *rolling plains,* which occupy 65 percent of the surface. These plains contain several large troughs which are of possible tectonic origin. Several circular features with shallow flat floors and bright rims have been identified. These may be impact craters. Several large plateaus are also present. The third type of Venusian landform is the *highlands.* These may be analogous to the large continental land masses on Earth. There are several Venusian "continents" that rise a couple of kilometers above the mean surface level, but these comprise only about 5 percent of the total surface area. The two most prominent of these features are Ishtar Terra and Aphrodite Terra. A prominent feature of Ishtar is a huge mountain named Maxwell Montes that towers eleven kilometers above the surrounding plain. Maxwell is two kilometers higher than Mount Everest on Earth and may actually be a giant shield volcano. Just to the east of Aphrodite is a giant rift valley known as Diana Chasma that is 900 kilometers long and 200 kilometers wide. It is 4.8 kilometers deep, much deeper than any canyon on Earth. Nearby is a large quasi-circular trough known as Artemis Chasma, which may be the remains of a giant impact crater.

The discovery of the high surface temperatures has effectively eliminated the possibility of any type of life on Venus. Surface photographs returned to Earth by the Soviet Venera landers show only a bleak landscape, apparently devoid of any life forms. The thick and dense carbon dioxide atmosphere protects the surface from ultraviolet light and the solar wind, but unfortunately traps so much heat that it renders the temperature far too high for water to exist in the liquid state. The crucial substrate for life is absent.

At one time Venus probably had enough water for an ocean to have covered much of the surface. During this early epoch, some primitive organisms could have evolved in the primeval Venusian seas. All of them, of course, would have perished at the time of the "runaway greenhouse" disaster which produced rapidly rising surface temperatures and eventually stripped Venus of much of its water. Perhaps the fossils of some of these early life forms are still present in the surface rocks. Venus may be a tragic world, one where life was snuffed out in its infancy.

Mars

Of all the other worlds in the solar system, Mars has for long been considered as the one with a surface environment most like Earth's and the one other planet most likely to harbor some sort of life. Mars is approximately one and a half times farther from the Sun than Earth. It is only about a tenth as massive as our planet and has a diameter of 6760 kilometers. The density is only 3.86 grams/cc, appreciably less than that of any of the inner three planets and more nearly equal to that of the Moon. The Martian interior is probably similar to that of Earth, with a metal-rich core and a silicate-rich mantle and crust. The central core, however, may be somewhat less extensive. There is only an exceedingly weak magnetic field, and there are no radiation belts.

The Phantom Canals. The first astronomical studies of Mars seemed to indicate that the planet had an atmosphere similar to Earth's, although in all probability appreciably cooler and dryer. Mars's equator is inclined at an angle of 25 degrees with respect to the plane of its orbit around the Sun, so it goes through seasons just as the Earth does. Astronomers could watch seasonally varying polar caps and could even see a seasonally changing color and texture that could be indicative of dense vegetation, which advanced in summer and retreated in winter. Some observers even imagined that they saw a series of canals on the surface, evidence perhaps of an intelligent civilization on Mars. Some more fanciful investigators even proposed that the sudden discovery of two Martian moons in 1877 after years of fruitless search was evidence that these two moons were actually huge artificial satellites launched only a few years earlier by an advanced Martian civilization!

This picture of Mars was accepted in many of its details by most astronomers up until the advent of the space age, although most doubted the existence of the canals. This view changed abruptly when the first spacecraft explorations of the planet were made in the 1960s. These studies showed that the Martian atmosphere is exceedingly thin, with a surface pressure of only 0.008 atmospheres. The atmosphere is largely carbon dioxide, with lesser amounts of argon and oxygen. There are trace amounts of water vapor in the atmosphere which show seasonal variation. It is rather cold almost everywhere on Mars. Daytime temperatures near the equator can reach as high as 0 degrees Celsius, but it drops down to −85 degrees Celsius at night. Near the poles it is a lot colder still, never getting any warmer than −70 degrees Celsius.

Large impact craters cover much of the Martian surface, but they are far more numerous in the southern hemisphere than in the northern. In addition, there are some large basinlike areas such as Hellas, Argyre, and Isidis, having dimensions of the order of 1000 to 2000 kilometers. These basins and craters were created by the impacts of asteroid-sized bodies over 4 billion years ago. The northern hemisphere is covered with large numbers of extinct volcanoes, most much larger than any on Earth. The most spectacular volcanoes are located in the Tharsis region, where there are three large ones (Arsia, Pavonis, and Ascraeus) lying in a row near the equator. A thousand kilometers away lies the huge Olympus Mons, 500 to 600 kilometers across and towering 23 kilometers above the surrounding plain. Judging from the relative scarcity of craters on their slopes, the three Tharsis volcanoes probably range in age from 400 to 800 million years. Olympus Mons may be only 200 million years ago. Other volcanic features are a good deal older. The era of Martian volcanism began almost immediately after the period of intense meteorite bombardment came to an end and continued until quite recent times. There is an enormous complex of canyons to the east of Tharsis, collectively named Valles Marineris. It is 2700 kilometers long, and

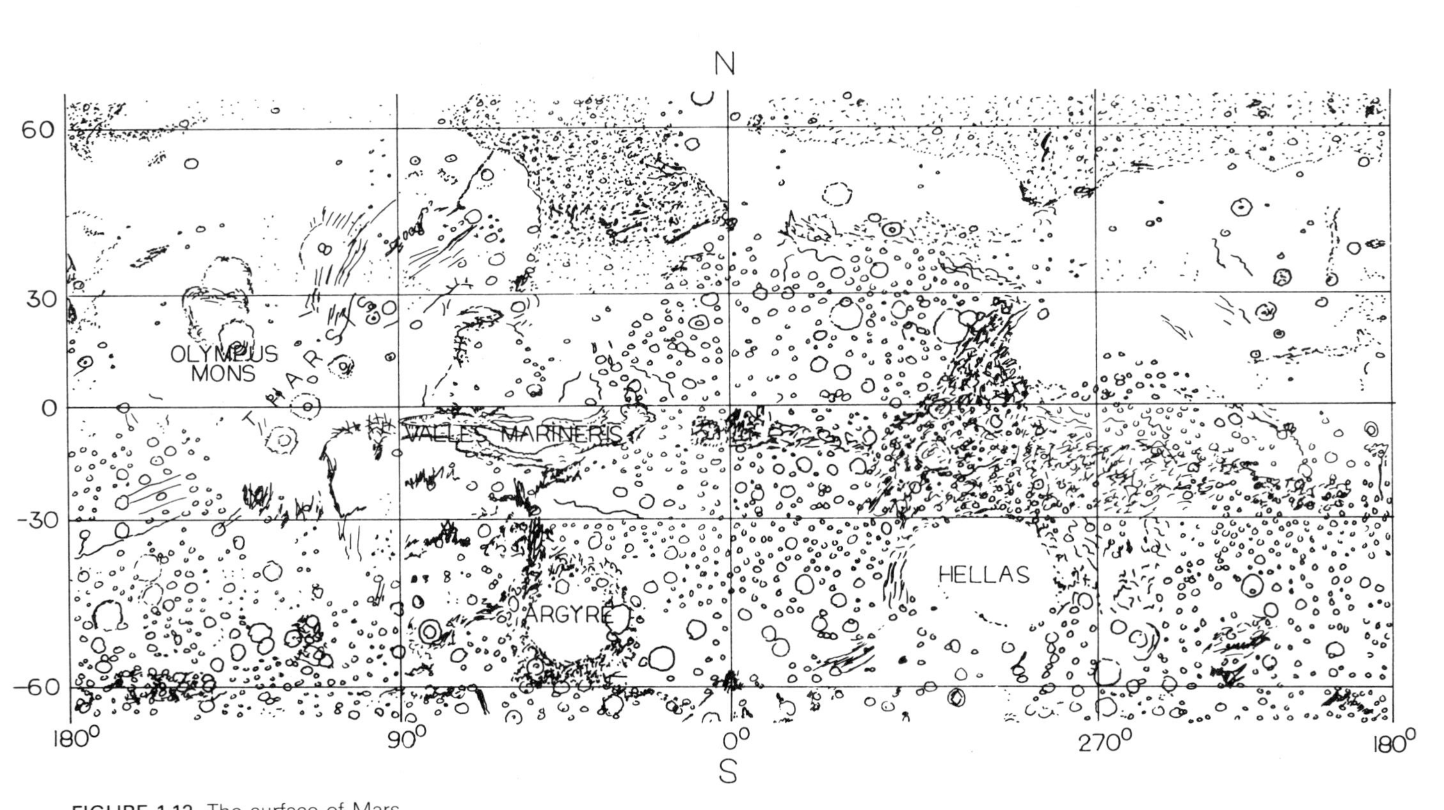

FIGURE 1.12. The surface of Mars.

individual canyons in the complex are up to 200 kilometers wide and 6 kilometers deep, easily dwarfing the Grand Canyon of Arizona on Earth.

The Runaway Icebox. There is no liquid water anywhere on Mars. What water there is on the planet is tied up either as water vapor in the atmosphere or as water ice in the soil and in the polar caps. Because of the low surface pressures, water cannot exist in the liquid state anywhere on Mars. The polar caps are made up of thick layers of water ice, but their seasonally variable component seems to be dry ice. The seasonally variable texture of Mars seems to be caused by water vapor freezing to the ground during the winter and evaporating during the summer rather than by any sort of biological activity. Seasonal planetwide dust storms may also play a role.

Although there is presently no liquid water on Mars, there is ample evidence for its presence in the past. There are numerous sinuous channels with tributary systems that appear to have been originally cut by running water. There are areas with complex braided patterns produced by the deposition of silt and rocky debris from running water. There are even a few teardrop-shaped "islands" that were produced by water flowing around obstacles. Perhaps Mars at one time had a much denser atmosphere, one that could trap enough heat near the surface to permit liquid water to exist. Extensive episodes of rainfall may have taken place, and rivers, lakes, and oceans of water may have covered much of the surface. However, because of the small size of the planet, most of this early atmosphere was lost to outer space. This caused the surface temperature to drop. As more and more of the Martian atmosphere disappeared, the cooling surface temperatures forced the liquid water to freeze solid into the surface soil. It remains there today, perpetually frozen into solid ice. The process by which all of a planet's water freezes into solid ice has come to be known as "runaway glaciation." It is essentially the inverse of the "runaway greenhouse" effect responsible for the high temperatures on Venus.

The Ambiguity of Viking. Any hopes for the presence of intelligent life on Mars or for any higher forms of life at all were dashed by the findings of the first unmanned spacecraft to explore the planet. Still, there might be some hardy microscopic organisms present in the Martian soil that had evolved at an earlier time when the climate was more hospitable. Two Viking spacecraft landed on Mars in 1976 to answer this question. The landers contained sophisticated miniature biochemical laboratories designed to test Martian soil samples for any kind of biological activity similar to that found on Earth. One experiment tested Martian soil for its ability to metabolize a "soup" of organic molecules. Another experiment attempted to find if Martian soil could metabolize a mixture of organic compounds that had been labeled with radioactive carbon-14. A third experiment attempted to determine if Martian soil was capable of photosynthesis by subjecting it to light and labeled carbon dioxide. In addition, there was a gas chromatograph mass spectrometer aboard to test Martian soil for the presence of organic compounds.

The Martian soil was checked for organic compounds by heating a sample to 500 degrees Celsius. Any volatile materials in the soil are driven off by the heat and sent to a gas chromatograph where their

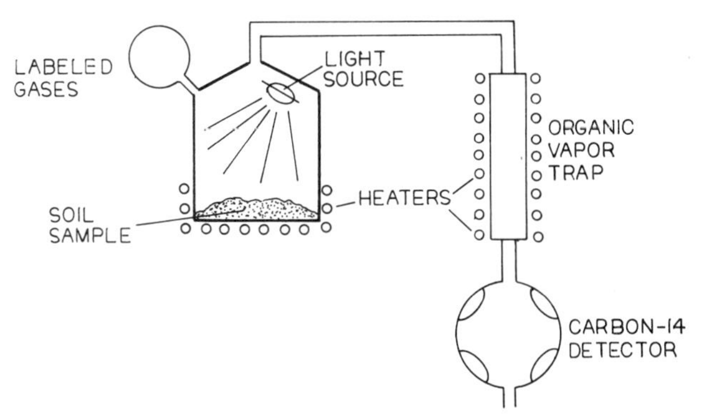

FIGURE 1.13. Schematic arrangement of the three Viking tests for life on Mars.

molecular weights are measured. To everyone's surprise, no gases other than carbon dioxide and water vapor were driven from the soil samples. No organic molecules were detected. The Viking gas chromatographs have been able to measure significant amounts of organic material in desert sand, in Antarctic soil, even in Moon dust. The total absence of organic molecules on Mars is at first sight perplexing, as there should be a continual influx of organic-rich carbonaceous chondrite meteorites. In addition, there should be a continuous creation of organic material via ultraviolet photolysis of water vapor and carbon dioxide in the atmosphere. However, a lot of ultraviolet sunlight can reach the Martian surface, apparently enough to completely destroy any organic molecules. In addition, it has been proposed that the ultraviolet light creates highly reactive inorganic peroxides at the surface. These quickly decompose any organic molecules that are present. Without organic molecules, the prospects for Martian life are exceedingly remote.

In the gas exchange experiment, Martian soil was tested for microorganisms that can absorb nutrient materials and give off gaseous by-products. The experiment was performed in two stages. First, a small volume of nutrient solution was added in such a way as to humidify the chamber without actually touching the soil sample. In the second stage, the sample was entirely wetted with nutrient and incubated for nine months. Any gases emitted by the sample were analyzed by a gas chromatograph. An unexpected thing happened. Immediately after the soil was first humidified, an intense burst of carbon dioxide and oxygen gas was detected. Was this the first evidence of a Martian biochemistry? The initial enthusiasm generated by this apparently positive result was quickly dampened, as the rate of gas release soon began to slow. When the sample was subsequently wetted with nutrient, the rate of carbon dioxide release slowed down and the oxygen completely disappeared. Everything of interest seemed to happen in the initial, humid stage of the experiment. The reaction was probably chemical, not biological. The carbon dioxide probably came from adsorbed atmospheric gases driven out of the soil by the heat. The oxygen must have been released by some sort of chemical reaction unique to Mars; such oxygen generation does not take place when terrestrial soils are incubated in the dark. It has been proposed that the source of the oxygen generation is the peroxides or superoxides presumed to be present on the Martian surface. The rapid disappearance of the oxygen upon complete wetting is thought to have been caused by its reaction with the ascorbic acid in the nutrient broth. Whatever the oxygen source, it seems to be unaffected by heating to 145 degrees Celsius for three hours. This should have killed any oxygen-generating organisms present, especially ones that had originally evolved on the cold surface of Mars.

The labeled-release experiment tested the ability of Martian soil to metabolize an aqueous nutrient broth of simple organic molecules that had been enriched with radioactive carbon-14. A small amount of radioactive nutrient was added to the soil sample, small enough to wet only part of the soil but large enough to humidify the entire chamber. The mixture was incubated for a couple of weeks. A subsequent nutrient injection brought the entire soil sample into contact with the broth. The gases emitted from the soil during the entire process were analyzed for their radioactivity. Any radioactive gases released would imply the existence of a process capable of splitting the organic molecules in the broth into simpler molecules. The results were similar to those obtained in the gas exchange experiment. There was an initial burst of radioactive gas immediately after the nutrient was added. This gas was probably carbon dioxide, and its evolution tapered off after a day or so. The subsequent complete wetting evolved no additional radioactive gas; in fact, the amount of radioactive gas in the chamber actually began to decrease. These results appear to have a chemical rather than a biological explanation. Peroxides present in the Martian soil could have attacked the formic acid in the broth, giving off carbon dioxide and water vapor in the process. The decrease in the radioactive gas concentration that took place after the complete wetting was perhaps caused by the dissolving of some of the radioactive carbon dioxide in the aqueous broth. However, there are some aspects of this experiment that leave biological activity open as at least a remote possibility. When one of the Martian soil samples was preheated to 160 degrees Celsius for three hours, the initial release of radioactive gas was abolished. Since such treatment also destroys terrestrial biological activity, a few workers have proposed that this result could possibly be indicative of some sort of biological activity on Mars.

The pyrolytic-release experiment was designed to test the ability of Martian soil to incorporate radioactive carbon monoxide or carbon dioxide into more complex molecules, either via photosynthesis or by some sort of dark biochemical process. A soil sample was sealed into a chamber, and radioactive CO and CO_2 gases were added. A xenon arc lamp irradiated the sample for five days. The lamp was then turned off, and the atmosphere was pumped away. The sample was then heated to a temperature high enough to pyrolyze into small volatile fragments any organic compounds that had been synthesized. These fragments were then swept into a column which was designed to trap organic molecules but pass carbon monoxide or carbon dioxide. The column was then heated in an oxidizing ambient; any organic molecules trapped were oxidized to CO_2 and carried into a radioactive counter. Any radioactivity measured would be an indication of the existence of something in Martian soil capable of synthesizing complex molecules out of carbon dioxide gases. Seven out of nine tests gave positive results, including one sample incubated in the dark. The amount of carbon fixed was, however, rather small, only about a tenth of that produced in similar tests on Antarctic soil. Does this mean that there is some sort of low-level biological activity taking place on Mars? This is thought to be improbable, since the reaction was much less sensitive to heat than would be expected for a biological process. The preheating of samples to 90 degrees Celsius for two hours had absolutely no effect on the reaction. A 175-degree Celsius heat treatment for three hours reduced the reaction by 90 percent but did not completely abolish it. Any terrestrial organisms would have been killed.

Although there is still a good deal of disagreement about the meaning of the results of the pyrolytic-release experiment, it is possible that some sort of nonbiological process within the Martian soil was able to take the carbon dioxide or carbon monoxide added during the Viking experiment and use it to synthesize more complex organic molecules. The inorganic peroxides presumed to be in the Martian soil could have combined with carbon monoxide in the presence of light during the experiment, giving a small amount of formic acid. The controversy is still continuing.

Does Mars have life or doesn't it? The bulk of the evidence seems to indicate a negative answer. Still, there are those few suggestive results that at first sight seem to have only a biological explanation. It is possible that there could be a few hardy Martian bacteria (perhaps only a dozen per gram) in the soil, enough to have given a weakly positive response to the three biochemical tests, but still without enough organic material in their tiny bodies to be measurable during the gas chromatography tests. On Earth, even the least biologically active soil has enough dead organic matter intermixed to give a positive result during the gas chromatography measurements, but the few Martian bacteria present might be such efficient scavengers that they effectively absorb all the available dead organic material into their bodies, leaving virtually none free in the soil.

Jupiter

Jupiter is by far the most massive of the planets. It lies in an orbit that is 5.2 times farther from the Sun than the Earth. Jupiter's mass is 318 times that of Earth, but still only 0.001 that of the Sun. Jupiter is particularly interesting because, unlike the inner four planets, it is made up largely of liquids and gases. Jupiter is sometimes called "the star that failed," as it has a composition nearly the same as the Sun's but is not nearly large enough to maintain the high internal temperatures and pressures needed to sustain thermonuclear fusion.

Jupiter appears in the telescope as a slightly flattened yellowish disk with alternating light and dark bands running parallel to its equator. Different parts of the Jovian disk appear to be rotating at slightly different velocities, indicative of a largely fluid interior. The equatorial diameter of Jupiter is 143,100 kilometers, and the average density is only 1.4 gm/cc. Theoretical calculations indicate that no elements other than hydrogen or helium could possibly give Jupiter such a low density under the enormous gravitational compression in the interior of so massive a body. The interior of Jupiter is at such high pressures and temperatures that the hydrogen and helium are in the liquid rather than the gaseous state. At a depth of 30,000 kilometers into the planet, hydrogen breaks up into a mixture of protons and electrons. This "liquid-metallic" phase of hydrogen probably extends nearly all the way to the center of the planet. Detailed models of the Jovian interior indicate that there is probably a dense central core of approximately 28 Earth masses. It may be rich in metals and silicates.

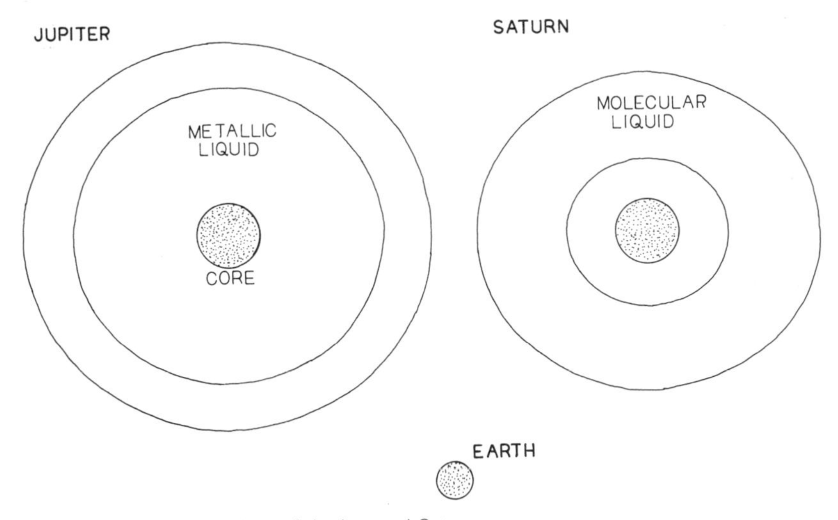

FIGURE 1.14. The interiors of Jupiter and Saturn.

Electrical currents within the liquid metallic interior maintain an enormous magnetic field around the planet, much more intense than Earth's field. This magnetic field traps charged particles in radiation belts that look a lot like Earth's Van Allen belts. The radiation intensity within the Jovian magnetosphere is thousands of times more intense than it is within the terrestrial radiation belts. The Pioneer 11 spacecraft passed within 42,000 kilometers of Jupiter in 1974 and received a radiation dose a hundred times the amount which would have been fatal to a human being. The radiation was so intense that some of the more delicate electronic components aboard the craft were permanently damaged.

Jovian Air. The very outermost regions of Jupiter that are visible from Earth consist of an enormously dense and thick atmosphere. This atmosphere is 81 percent hydrogen and 19 percent helium, although trace amounts of methane, ammonia, water vapor, phosphine, germane, acetylene, and ethane are also present. The upper reaches of the atmosphere are rather cold (−160 degrees Celsius), but it gets steadily warmer as one descends. The pressure also rises with depth. It is not at all certain whether or not there is any kind of surface beneath the atmosphere. Some models predict that the gaseous atmosphere continuously blends into the more dense liquid hydrogen interior of the planet, with no clear and distinct interface existing between the two. Other models predict a definite liquid hydrogen surface underneath Jupiter's atmosphere, with a sharp density increase at a depth of 1000 kilometers where the temperature is 2000 degrees Celsius and the pressure is 500 atmospheres. A liquid

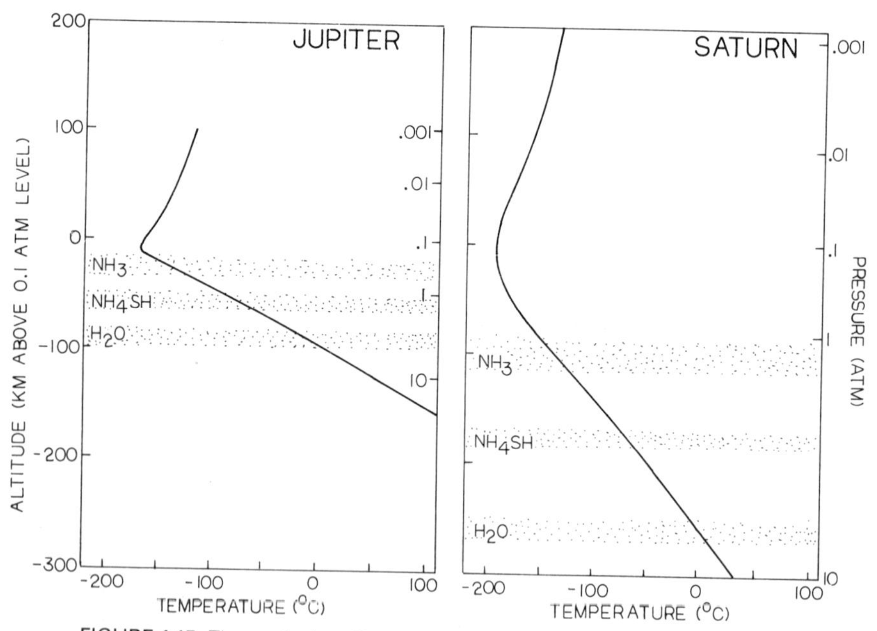

FIGURE 1.15. The vertical profile of the atmospheres of Jupiter and Saturn. From Moore and Hunt (1983). Copyright 1983, Mitchell Beazley Publishers, Ltd. Used by permission.

hydrogen "ocean" lies at this level, locked in perpetual darkness under crushing pressures and broiling temperatures.

There is a complex layer of thick clouds in the upper atmosphere. The composition of the clouds is uncertain, but it is thought that they are largely condensed ammonia, water, and ammonium hydrosulfide vapors. The Jovian upper atmosphere cloud deck exhibits a rich pattern of colors. The substances responsible for the colors are uncertain, as all of the suspected cloud condensates are white. It is possible that the chromophores are organic molecules, created by the ultraviolet photolysis of the methane, ammonia, and water vapor in the upper Jovian atmosphere.

Jovian Life? At first sight, Jupiter appears to be a poor candidate for the presence of life. There is no solid surface, and the temperatures and pressures deep in the atmosphere are so enormous that no complex organic molecules could survive. No sunlight reaches down below the cloud deck, making any sort of photosynthesis impossible. However, at higher altitudes pressures and temperatures could be low enough for some organic molecules to remain stable. The Jovian atmosphere is certainly rich enough in water vapor, ammonia, and methane, the primary precursors in the formation of more complex organic molecules, and lightning bolts might have synthesized a rich organic "mist" in the upper Jovian air which gives the clouds their striking pattern of colors. It is even possible that living organisms are present, perpetually floating in the dense upper atmosphere. Carl Sagan and Edwin Salpeter of Cornell University have proposed that there could be an airborne Jovian ecology that has three distinct types of organisms, termed *sinkers, floaters,* and *hunters.* The sinkers are born in the upper atmosphere, where they feed on mists of organic molecules floating high in the Jovian clouds. They slowly drift downward as time passes, eventually to be devoured by the floaters, which live at lower levels. The floaters then act as prey for the hunters, which are pictured as gaseous, balloonlike creatures several kilometers in size that move to and fro via a sort of jet propulsion. A hunter could be large enough to appear in the television pictures returned by the Voyager flypast spacecraft, and these photographs should be carefully scanned for evidence of such a creature. The Great Red Spot, with its prominent coloration, may be a particularly favorable spot to look.

Although the prospects for Jovian life are exciting, several serious objections can be raised. For one, there is very little liquid water in the upper Jovian atmosphere to act as a substrate for biochemical reactions. The Jovian atmosphere is considerably drier than would be expected for a mixture of vapors similar in composition to the Sun. What little water there is in the vapor state or perhaps in the form of ice crystals at the top of the Jovian cloud system, although it is possible that there are some liquid droplets farther down in the atmosphere. Another problem is that Jovian weather is so violent that any organisms formed in the upper atmosphere would soon be carried downward by vertical air currents, where they would instantly be fried by the high temperatures and pressures.

Saturn

Much of what has been said about Jupiter is equally valid for the giant ringed planet Saturn. Saturn is approximately twice as far from the Sun as Jupiter. The planet is a dense gaseous ball made up of hydrogen and helium rather than metals or silicates. The ring system is a gigantic sheet of water ice particles, each of which travels around Saturn in its own separate orbit.

Apart from the rings, Saturn is similar in size and overall appearance to Jupiter. The internal structure is similar, with an outer layer of liquid hydrogen and an inner region of liquid metallic hydrogen. There appears to be a dense metal-rich central core of approximately 19 Earth masses.

Saturn's atmosphere is a lot like Jupiter's, being primarily hydrogen and helium in composition. Trace amounts of ammonia and methane are also present, but as yet no water vapor has been found. The upper atmosphere is somewhat colder than Jupiter's, although both the pressure and temperature increase with depth until the atmosphere merges with the liquid hydrogen below. There are dense clouds in the upper atmosphere. They may consist of condensed ammonia, ammonia hydrosulfide, and water vapors. There are also patterns of colors in the upper cloud deck, but the colors are not nearly so rich as they are in the Jovian atmosphere.

Saturn also has an extensive magnetosphere, driven by persistent electrical currents within the liquid metallic hydrogen interior. There is a system of radiation belts associated with the magnetic field. However, Saturn's ring system absorbs a large fraction of these charged particles, so the radiation intensity is a lot weaker than it is in the Jovian magnetosphere. However, it is still strong enough to be fatal to any astronauts foolish enough to fly through it.

The comments about the possibility of life in the Jovian atmosphere are equally valid for Saturn. Saturn's upper atmosphere is even drier than Jupiter's, although there may be appreciable amounts of water at lower levels. The blander appearance of the Saturnian atmosphere may mean that there are fewer organic molecules to act as building blocks for organic molecules. As a result, life in the Saturnian atmosphere is very unlikely.

Uranus and Neptune

The two outer planets Uranus and Neptune are virtual twins. Uranus is twice as far from the Sun as Saturn, and Neptune is 10 AU farther out still, so these worlds must be exceedingly cold. They are approximately 50,000 kilometers in diameter, which makes them intermediate in size between the small inner terrestrial planets and the outer gas giants Jupiter and Saturn. These worlds seem to be largely fluid and gaseous in composition rather than solid. However, they cannot be entirely or even largely made up of hydrogen. The dominant components are most likely water, ammonia, and methane. Models of their interiors indicate that there is probably an outer liquid hydrogen "ocean" much like Jupiter's, but there is a mantle of "slush" rich in ices of water, methane, and ammonia in the interior which actually comprises at least two-thirds of the mass. Neither of these worlds is believed to be large enough to sustain any liquid metallic hydrogen in its interior. Both of these worlds probably have rocky and/or metallic cores that are on the order of 3 to 4 Earth masses in size.

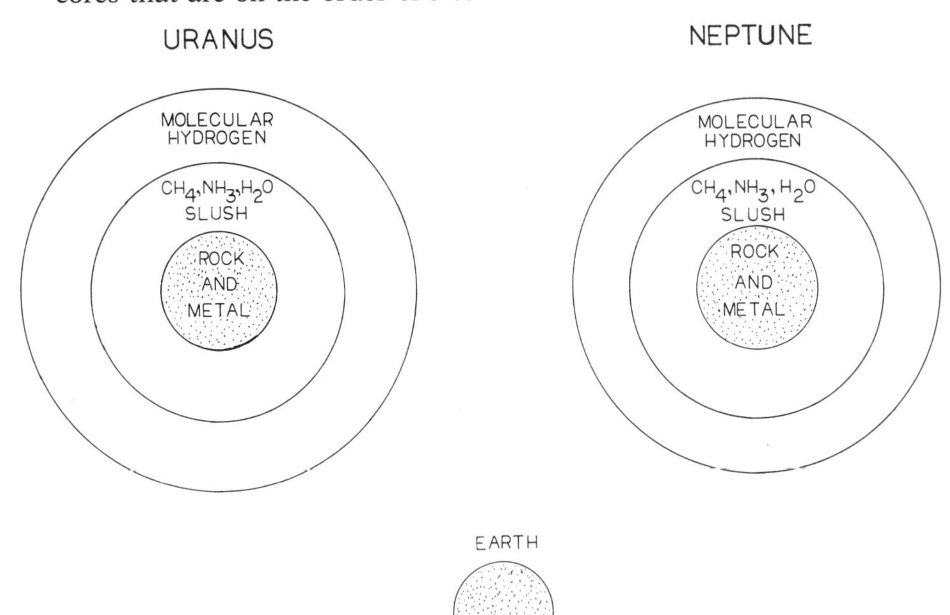

FIGURE 1.16. The interiors of Uranus and Neptune.

The atmospheres of these worlds are very largely unknown quantities, as no spacecraft has yet flown past them. The dominant gases are hydrogen and helium, intermixed with an appreciable amount of methane. No water vapor or ammonia have been found, although appreciable amounts of both may be present at lower levels. The upper reaches of their atmospheres are exceedingly cold, less than −200 degrees Celsius. However, both the pressure and temperature undoubtedly increase with depth until the atmosphere merges with the liquid hydrogen "ocean."

The comments about the possibility of life on either Jupiter or Saturn are equally valid for Uranus or Neptune. Neither world has a solid surface, and the atmospheres are so dense that the lower levels are in perpetual darkness. The upper atmosphere is so cold that water, if it is present, can only be in the solid state as tiny ice crystals in the clouds. The prospects for life on either Uranus or Neptune are exceedingly remote.

Pluto

The outermost planet, Pluto, is in a rather eccentric orbit that averages 40 AU from the Sun. The orbit is so eccentric that this planet actually passes inside the orbit of Neptune during part of its journey around the Sun.

Pluto is an exceedingly tiny world, with a diameter of only 3000 to 3600 kilometers. The mass is only 0.1 percent of that of Earth. The density is exceedingly small, less than 1 gm/cc. Such a low density must mean that Pluto is made up largely of frozen volatiles such as methane, ammonia, or water rather than silicate rocks or heavy metals. The spectrum of solid methane has been detected in the light reflected from the planet, and it is possible that the planet is actually made up almost entirely of frozen methane.

The effective temperature of the bright side of Pluto is −230 degrees Celsius. In spite of its small size and cold temperatures, there is the distinct possibility that Pluto has an atmosphere. Some infrared reflectance spectra seem to indicate that there is a thin methane atmosphere, with a surface pressure as high as 0.05 atmospheres. The atmosphere is probably not so thick that the surface of Pluto cannot be seen from outer space.

The cold temperatures on the surface of Pluto rule out the presence of any sort of life as we understand it. There may be appreciable amounts of methane, ammonia, and water present to act as a substrate for life, but these materials are largely in the solid state at these cold temperatures. The only water present on the surface is in the form of solid ice. Ultraviolet light and solar wind irradiation should be negligible at these extreme solar distances, but sunlight is so feeble that photosynthesis could not possibly function.

Jupiter's Galilean Satellites

Jupiter has sixteen known moons. All but four are exceedingly small bodies that can be seen only in the largest telescopes. The four exceptions, termed the Galilean satellites, are large enough to be considered significant worlds in their own right. Do any of them have any prospect of having life?

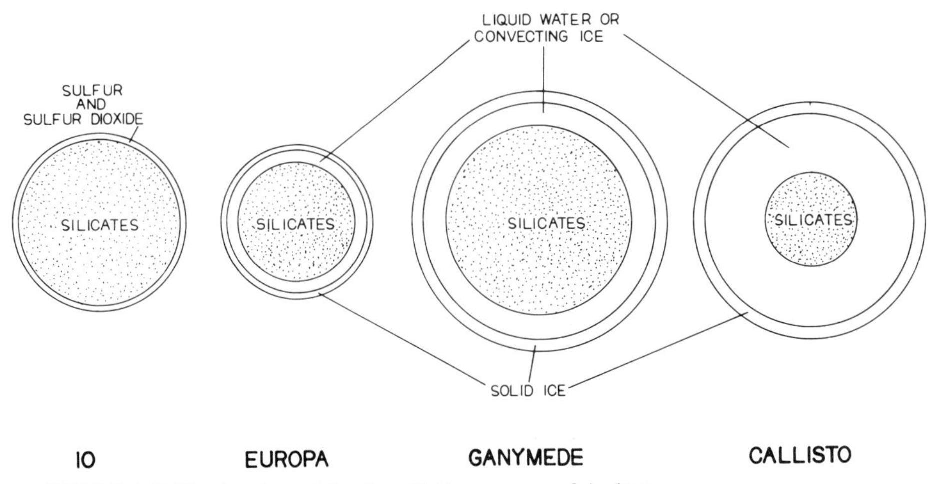

FIGURE 1.17. The interiors of the four Galilean moons of Jupiter.

Io. The innermost Jovian moon, Io, is perhaps the most interesting world in the entire solar system. Io is a bit larger than Earth's Moon and has no significant atmosphere. It is a brightly orange-colored sphere with irregular white splotches. There are no impact craters, but there are a large number of odd-looking sinkholes scattered over the surface. To everyone's surprise, these sinkholes turned out to be the vents of active volcanoes.

Ionian volcanism is entirely different from terrestrial volcanism. Io's density is consistent with the presence of a rocky, metal-rich inner core and an outer crust of sodium and potassium salts that is rich in sulfur-containing compounds. The upper mantle of Io is probably entirely molten, with pools of liquid SO_2 and vast underground rivers of molten elemental sulfur. Molten sulfur is heated by thermal energy from the interior and is forced upward to the surface. When it comes into contact with the liquid SO_2, the SO_2 is vaporized and forced upward through fissures until it vents to the surface. The SO_2 vapor quickly condenses into a solid when it is exposed to the vacuum of outer space, eventually falling to the ground and covering the surface with a thick blanket of frozen "snow." The excess molten sulfur from the interior then flows upward and spills out of the volcano and covers the surrounding surface. Io is the most geologically active world in the entire solar system; there is enough volcanism to renew the surface once every few million years.

Could life have risen on such an active world? There seems to be little if any water present on the surface and probably not much in the interior. The critical substrate for life is absent. In addition, most places on Io are rather cold. Temperatures average about −145 degrees Celsius during the day, although it may get as hot as 300 degrees Celsius near the active volcanoes. Nighttime temperatures get as low as −180 degrees Celsius. There is no significant atmosphere, so the surface is exposed to the Sun's dangerous ultraviolet rays. Perhaps the most important hazard to life on Io is the intense radiation within the Jovian magnetosphere. Io orbits in the most intense region of the radiation belts. Any organisms on the surface would be killed.

Europa. The second Galilean moon, Europa, is slightly smaller than Io and is somewhat less dense. It looks a lot different from Io, with a light brown surface interspersed with white patches. Much of the terrain consists of interlocked depressions and mesas a few kilometers in size. There are few impact craters, but no evidence of any volcanism either now or in the distant past. Europa is unique in having a surface crisscrossed with a vast network of narrow dark stripes. The lines look a lot like some of the drawings of the notorious Martian canals published early in the twentieth century.

The reflectance spectrum of Europa shows evidence of the presence of large amounts of water ice. The density is consistent with as much as 20 percent of the mass being water ice, with the rest being metals and silicates. The absence of impact craters on Europa must mean that the surface is so soft and pliable that any large craters disappear in only a few million years, an incredibly short time on the scale of the history of the solar system. The underlying icy crust of Europa must be significantly warmer than the maximum daytime surface termperature for the topographical relief to relax so rapidly. There might be enough heat within the interior to melt some of the deeper icy crust of Europa, enough perhaps to have formed a 100-kilometer thick mantle of liquid water starting 40 kilometers under the outer layer of ice.

This underground Europan "ocean" may be the only hope for extraterrestrial life in the solar system. When the solid outer crust cracks, some water from the mantle may spew forth from the cracks to settle upon the surface. This water will freeze solid in about three to five years. Sunlight may penetrate into the lower liquid mantle at points near the cracks, and it is at least conceivable that the regions of the Europan liquid water mantle near these cracks could be nearly as warm and well lit as some of the perpetually frozen lakes in Antarctica. These frozen lakes do support life, and it is possible, although not certain, that some sort of primitive organisms could be living in the Europan oceans under the protective blanket of the solid outer icy crust, where they would be shielded from the intense radiation of the Jovian magnetosphere.

Ganymede and Callisto. The outer two moons, Ganymede and Callisto, are entirely different. Ganymede has a diameter of 5260 kilometers, which makes it slightly larger than the planet Mercury. Callisto is somewhat smaller, with a diameter of 4900 kilometers. Both of these moons are significantly less dense than Io or Europa. They are approximately 50 percent water ice in composition, with an inner core rich in silicates. Neither world has even a trace of an atmosphere. Maximum daytime temperatures are −110 degrees Celsius.

Superficially, Ganymede and Callisto both resemble the Moon, with large expanses of densely cratered terrain. Ganymede's surface consists of a set of dark, densely cratered regions that are roughly polygonal in shape. These regions are separated from each other by broad stripes of brighter terrain that is generally less densely cratered. Many of the broad bright stripes have intricate sets of grooves, almost as if a giant rake had been passed over the surface. The darker regions consist of ice intermixed with silicate dust and rocky debris, but the brighter grooved terrain is probably nearly pure ice. Some sort of tectonic process took place very early in Ganymede's history, but all geological activity must have ceased about 3.5 billion years ago when the icy crust froze solid. Callisto is even more densely cratered than Ganymede and has no bright grooves. Its surface must be at least 4 billion years old.

It is probable that Ganymede and Callisto have significant amounts of liquid water deep within their interiors, so is it possible that they could have life? The densely cratered surfaces of these worlds indicate that their outer icy crusts are significantly thicker and less pliable than the Europan crust. Any underground ocean still existing beneath the crust must be so far below the surface that no sunlight could possibly penetrate. The prospect of any life at all inside these worlds is exceedingly remote.

Saturn's Giant Moon Titan

Saturn has seventeen known moons, more than any other planet. Some of them are rather substantial bodies with diameters of 400 kilometers or more. The moons are exceedingly frigid bodies largely made up of water ice. All but one are completely without atmospheres. The lone exception, Titan, has an atmosphere denser than Earth's and may be of substantial biological interest.

Titan has a diameter of 5150 kilometers, which makes it just a bit smaller than Ganymede. The density is consistent with Titan being a 50:50 mixture of rock and water ice. The dominant component of the atmosphere is nitrogen, with about 1 percent methane and 2000 parts per million hydrogen. As much as 10 percent of the atmosphere may be argon. There are trace amounts of such organic gases as acetylene, ethane, propane, ethylene, diacetylene, and hydrogen cyanide. The surface pressure may be as high as 1.5 atmospheres. The temperature at the surface appears to be rather cold, about −180 degrees Celsius. No water vapor could be expected in the Titanian atmosphere at such cold temperatures, and none seems to be there. However, there are almost certainly ample amounts of water permanently frozen solid to the surface.

There is a thick layer of reddish aerosols hovering in a layer about 200 kilometers above the ground that perpetually obscures the surface of Titan from view. The composition of the aerosols is uncertain, but it seems probable that they are solid particles rich in organic polymers. The upper atmosphere exhibits a rich pattern of photochemistry, with higher molecular-weight organic gases such as ethane and acetylene having been created by the ultraviolet photolysis of methane. Higher molecular-weight hydrocarbons may also have been formed, and these may account for the reddish color of the aerosol layer. There may be a steady rain of complex organic hydrocarbons falling onto the surface from the upper atmosphere, and the surface of Titan could actually be covered with a thick layer of oil!

Because of the aerosol layer, next to nothing in known about the surface of Titan. As a result, speculation can roam fairly freely. The surface temperature is quite near the "triple point" of methane, the temperature at which methane coexists in the solid, liquid, and vapor phases. There could be a Titan-wide methane "ocean" at the surface or even a solid methane "crust," although it is probably not cold enough for lakes or oceans of liquid nitrogen to form. Another possibility is that there is a global ocean of ethane at the surface, formed via an ethane "rain" pouring down from the upper aerosol layer. Alternatively, there could be an ocean of liquid ammonia at the surface as deep as 100 kilometers. The surface might even consist of a

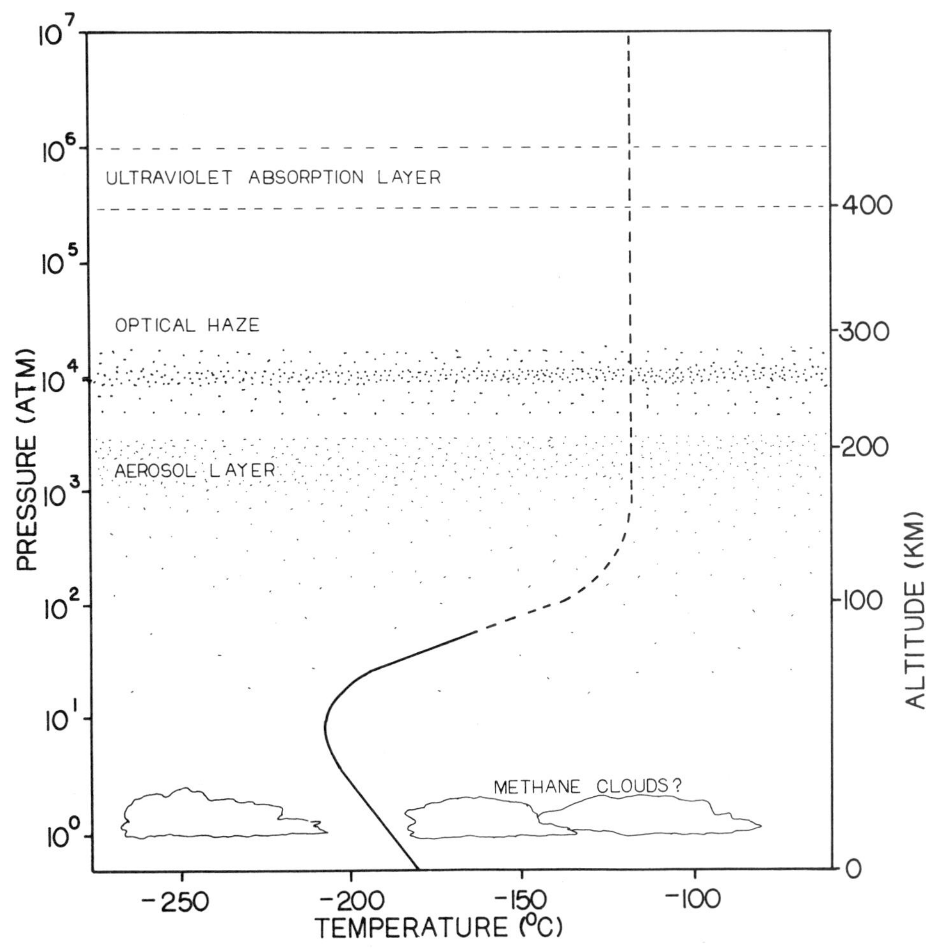

FIGURE 1.18. The vertical profile of the atmosphere of Titan. From B. A. Smith et al. (1981). Copyright 1981 by the American Association for the Advancement of Science.

mantle of ammonia-methane compounds that supports an *asthenosphere* made up of a liquid water-ammonia "slurry" topped off by an outer crust of water ice. Finally, there could conceivably be a thick "blanket" of hydrocarbons, amino acids, nitrogenous bases, and other such organic molecules, formed via a steady downpour of organic material from the upper atmosphere. The Titan landscape may be a surrealistic image of methane rivers cutting their way through frozen hydrocarbon canyons.

The stuff of life may be in abundant supply on the surface of Titan, but could life itself ever have formed on such a world? The cold surface temperature on Titan may rule out any sort of biological activity. Biochemical reactions (at least in terrestrial organisms) would be impossibly slow at such low temperatures. It is nevertheless conceivable that there exist living beings with metabolic processes that function at exceedingly slow rates. Perhaps there are conscious beings on Titan that live and think so slowly that they endure for thousands or millions of years. We can only speculate until a landing spacecraft reaches the surface. The atmosphere should filter out much of he ultraviolet light and particle radiation that would otherwise strike the surface of Titan, so any organisms would be well-protected from Saturn's intense radiation belts. However, the dense clouds block out so much sunlight that the surface of Titan is probably never any brighter than it is on a moonlit night on Earth. Photosynthesis under such circumstances may be impossible. Another problem is the apparent lack of liquid water. If there is any water on Titan's surface, it is certain to be frozen solid. Titan may be an organic chemist's dream, but it is in all probability devoid of life.

MINIMAL CONDITIONS FOR LIFE

We now summarize what the recent exploration of the solar system has taught us about the origin and evolution of living beings. We have discovered that the prospects for life on any planet in the solar system besides Earth are rather bleak. Hope for life on the most likely candidate, the planet Mars, appears to be fading. It now seems reasonable to conclude that an environment similar to Earth's is an absolute requirement for the development and maintenance of life. The absence of one or more of these essential conditions on a given planet or moon will prevent the formation of any sort of living organisms. These minimal conditions are summarized as follows:

Temperature

A moderate surface temperature is essential. Some primitive blue-green algae are able to survive and even thrive in near-boiling water at 85 degrees Celsius, and some animals have lived through temperatures as high as 50 degrees Celsius for short periods. However, all organisms are killed by boiling water. Bacteria cultured from high-temperature waters around sulfide-emitting chimneys ("black smokers") located on the ocean floor can actually survive and grow at temperatures of 250 degrees Celsius. Their survival is made possible by the high pressure (265 atmospheres) at the ocean bottom, which keeps the superheated water from boiling. Most plants cease to function below 0 degrees Celsius, but there are some animals that can survive brief exposures to temperatures as low as −30 degrees Celsius. At extremely low temperatures, biochemical reactions become impossibly slow and the water inside the cell body freezes, killing the organism. Although there are a few hardy exceptions, most complex plants and animals require temperatures between 0 and 30 degrees Celsius for survival and active growth. The most optimistic estimate of the range of survivable temperatures is probably −30 to 85 degrees Celsius.

Water

Liquid water is required for life. About 70 percent of the weight of living tissue is made up of water. This is, of course, a reflection of the fact that all life originated in the oceans, but it is also a reflection of the unique properties of water. Water is so familiar to us that we do not usually recognize how unusual and peculiar a substance it really is. Water exists in the liquid state at the moderate temperatures typically found on Earth. Most other simple molecules typically found in abundance in the solar system (carbon dioxide, ammonia, methane, hydrogen cyanide, hydrogen sulfide) are gaseous at these temperatures. Water is also unique in that it remains liquid over a far wider temperature range (from 0 to 100 degrees Celsius) than any other common substance. Water acts as an effective solvent for a large number of molecules and ions that are critically important in biological processes. Water has a very high heat of vaporization and a high specific heat, important properties in the regulation of the body temperatures of living creatures. The ionization products of water (H^+ and OH^-) are essential ingredients in a whole host of biologically significant processes. Last but perhaps far from least, liquid water has the somewhat unusual property of freezing from the top down rather than from the bottom up. A layer of ice covering the oceans at the north and south poles acts as an insulating "blanket" which prevents the oceans from freezing all the way to the bottom during the cold snaps which have occurred several times in the past. Water seems to be so ideally suited as the substrate for life that it is difficult to imagine any substitute.

Some have proposed that liquid ammonia could take the place of water as a substrate for life on planets and moons (such as Titan) that have extremely cold environments. Ammonia as well as water can participate in oxidation-reduction reactions and has some chemical properties at low temperatures that are analogous to those of liquid water at more elevated temperatures. Ammonia-based creatures might be able to survive at these cold temperatures by employing biochemical reactions that proceed at a very much slower rate than do those in water-based life on Earth. However, ammonia is liquid only over a relatively narrow temperature range, between −77 and −33 degrees Celsius. In addition, ammonia "ice" sinks in liquid ammonia. Ammonia oceans on these frigid worlds would probably freeze solid in rather quick order. As a result, the existence of ammonia-based life is unlikely.

The temperature range over which life is possible corresponds quite well to the range over which water can be maintained in the liquid state. The entire ecology of Earth is so utterly dependent on the presence of liquid water that it is difficult to imagine any sort of life possible without it.

Radiation Shielding

There must be some sort of shield provided against the harmful effects of ultraviolet rays from the Sun as well as from cosmic rays and the corrosive effects of the solar wind. On Earth, primitive life forms were initially protected from ultraviolet light by immersion in the oceans. At a later time, the Earth evolved an oxygen-rich atmosphere that in turn produced a protective ozone layer that could block out enough dangerous ultraviolet light so that life forms could move closer to the water's surface and then eventually onto dry land. Earth's dense atmosphere provides protection from highly energetic cosmic rays and screens out the dangerous X-rays and gamma rays that are emitted during solar flares and active parts of the sunspot cycle. The Earth's magnetic field prevents the solar wind from striking the atmosphere and disrupting the delicate ozone layer. It is difficult to imagine any effective shield against these harmful radiations other than deep oceans, a dense atmosphere, and a strong permanent magnetic field.

The Stuff of Life

An abundant supply of carbon, hydrogen, nitrogen, and oxygen atoms is essential for life. On Earth, these four atoms together make up over 99 percent of the matter found in living creatures. What are the unique properties of these particular elements that seem to make them so suitable for the formation of living matter? These four are the lightest elements that can form covalent bonds with each other (that is, bonds in which pairs of electrons are shared between adjacent atoms). Covalent bonds between these four elements are generally a good deal stronger and more stable than those formed between heavier elements. Unlike chemical bonds involving many of the other elements, covalent bonds within molecules made up of carbon, hydrogen, nitrogen, and oxygen are relatively stable against attack by oxidizing agents. Carbon, hydrogen, nitrogen, and oxygen happen tc be the only elements which can regularly form double and even triple covalent bonds to make compounds of exceedingly high stability. Carbon has the unique property of being able to bond directly to itself and to act as a framework for an incredible variety of complex molecules of widely differing shapes and sizes.

A few people have speculated that silicon might be able to play the role of carbon on some other worlds, since it too can combine with itself to form large complex molecules. Unfortunately, silicon-silicon bonds are readily attacked by oxygen, forming silicate compounds and such insoluble silicon polymers as quartz, the primary constituent of sand. A good deal of the Earth's crust is made up of silicate compounds, but nature has selected carbon for the formation of the covalent backbones of all the molecules essential for life. Carbon, oxygen, nitrogen, and hydrogen seem to be so uniquely suitable that it is difficult to imagine the existence of living creatures anywhere in the universe which do not use these elements.

Sunlight

Even though an atmosphere rich in gases containing carbon, nitrogen, oxygen, and hydrogen atoms is essential for life, this atmosphere must not be so dense nor the cloud cover so thick that sunlight cannot penetrate. Some sort of photosynthetic creation of essential nutrient and structural molecules is needed for the continual maintenance of life. Life probably began without photosynthesis, but it could not continue long without it. The pigment molecules, such as chlorophyll, that are essential for photosynthesis all absorb light in the visible and near-infrared band of the electromagnetic spectrum, between 3000 and 9000 angstroms in wavelength. Ultraviolet light of shorter wavelengths is so energetic that it disrupts the chemical bonds in the pigment molecules, whereas infrared light of longer wavelength is not energetic enough to intitiate any sort of photochemistry. An atmosphere transparent to light in this region of the spectrum is an absolute requirement for the continual existence of life.

Oxygen

An atmosphere rich in oxygen is probably a requirement for the development of higher multicellular organisms with brains and muscles, as well as for life capable of living outside the water on dry land. The presence of oxygen in the air may also be a prerequisite for the development of intelligence and the subsequent appearance of any sort of technology. Oxygen is certainly not a prerequisite for life itself; in fact, life on Earth appears to have originated in the total absence of molecular oxygen and probably could not have appeared at all if this gas had actually been present.

The prospects for life on the nine planets in the solar system, as well as on some of the larger moons, are summarized in Table 1.1. Apart from the Earth, all of the planets lack one or more of the essential characteristics required for the appearance and continual maintenance of life.

TABLE 1.1 Prospects for Life on the Planets and Moons of the Solar System

	Is the temperature within the habitable range?	*Is there water in the liquid state?*	*Is there a shield against harmful radiation?*	*Is there an adequate supply of C,H,N,O?*	*Can visible light penetrate?*	*Is there molecular oxygen?*
Mercury	No	No	No	No	Yes	No
Venus	only high in the atmosphere.	a small amount in upper air.	Yes	Yes	Yes	only a trace.
Earth	Yes	Yes	Yes	Yes	Yes	Yes
Moon	No	No	No	No	Yes	No
Mars	No	No	No	Yes	Yes	only a trace.
Jupiter	only high in the atmosphere.	perhaps in the clouds.	Yes	Yes	No	No
Io	No	No	No	No	Yes	No
Europa	perhaps under the ice crust	perhaps under the ice crust	perhaps under the ice crust	perhaps under the ice crust	perhaps in the region of ice cracks	No(?)
Ganymede	No	No	No	No	Yes	No
Callisto	No	No	No	No	Yes	No
Saturn	only high in the atmosphere.	perhaps in the clouds.	Yes	Yes	No	No
Titan	No	No	Yes	Yes	No	No
Uranus	only high in the atmosphere.	perhaps in the clouds.	Yes	Yes	No	No
Neptune	only high in the atmosphere.	perhaps in the clouds.	Yes	Yes	No	No
Pluto	No	No	No	No	Yes	No

CHAPTER TWO

LIFE AROUND OTHER STARS

The prospects for life on other planets in our solar system are exceedingly bleak. It now appears virtually certain that we will have to travel to other star systems if we are to have any success in finding other life forms, to say nothing of other technological civilizations. But do other stars actually have planets, or is our Sun somehow unique? Even if there are planets around other stars, what is the chance that some of these planets are suitable for the evolution of life? An attempt will be made here to give quantitative estimates of the various factors that affect the chances of finding other intelligent beings in the galaxy. These involve many variables, such as stellar spectral type, the probability of planetary formation, the nature of multiple star systems, and the types of planets capable of supporting life.

THE PROPER SPECTRAL TYPE

Our Star, The Sun

The star that is most familiar to us is of course the Sun. The Sun is made up of hot gases that assembled from a cloud of cosmic debris about 5 billion years ago. The elemental composition of the Sun's outermost surface can be measured by studying the spectrum of the light that is emitted. Ninety percent of all the atoms in the Sun are hydrogen, with the bulk of the remainder being helium. All of the other elements in the Sun are present in much lower abundances. For every 10,000 atoms of hydrogen, for example, there are seven atoms of oxygen, three of carbon, three of neon, and one of nitrogen. Elements of higher atomic weight are even more scarce; the relative abundance generally decreases with increasing atomic weight. Only 2 percent of the total solar mass is made up of these "heavy metals," which is an astronomer's shorthand for any element with an atomic weight greater than that of helium.

How has the Sun kept on shining so steadily for such a long time? The source of the Sun's energy is thermonuclear fusion reaction between the hydrogen nuclei deep in the interior. The mass of the Sun is so great that gravitational compression in the interior can produce extremely high densities and temperatures. The material in the solar core is so hot that it is entirely ionized, consisting largely of protons (hydrogen nuclei), alpha particles (helium nuclei), and electrons flying around at tremendous velocities. At such extreme temperatures, fusion reactions between hydrogen nuclei can take place, as indicated in Table 2.1. The net result of this chain of reactions is the

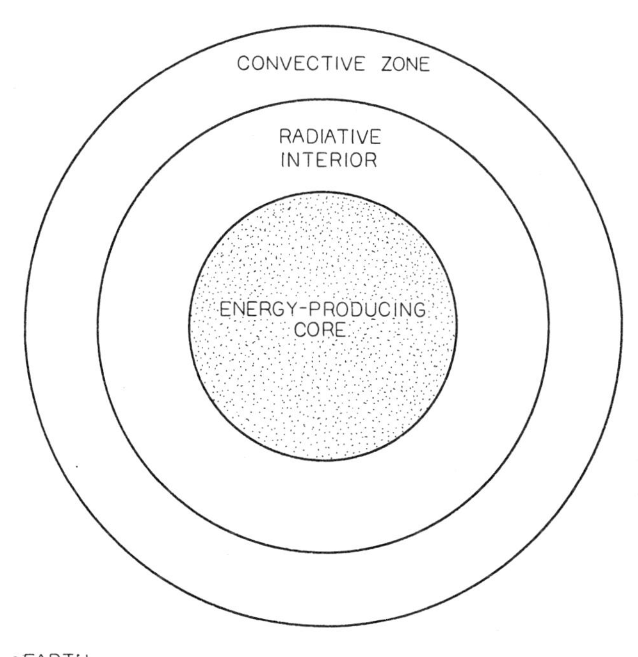

FIGURE 2.1. Interior of the Sun.

disappearance of four protons and two electrons and the production of a single helium-4 nucleus, eight gamma rays, and two neutrinos. The hydrogen in the solar core is gradually depleted, and the concentration of helium steadily builds up.

Compared to conventional chemical combustion, the energy released in these thermonuclear reactions is enormous. There is enough energy given off by the fusion of one gram of hydrogen to lift a weight of 10,000 tons to a height of 25,000 feet. The rate at which the Sun emits light energy is called the *luminosity*. This is equal to 3.9×10^{26} watts. The light emitted from the Sun closely matches the spectrum of an ideal "radiator" at a temperature of 5500 degrees Celsius. A body at that temperature emits its most intense radiation at a wavelength of 5000 angstroms, which corresponds to yellow light.

TABLE 2.1 Hydrogen Fusion Reactions in the Sun

Reactants		Products
${}^1H_1 + {}^1H_1$	$\longrightarrow$	${}^2H_1 + {}^0e_1$ + neutrino
${}^0e_1 + {}^0e_{-1}$	$\longrightarrow$	2 gamma rays
${}^2H_1 + {}^1H_1$	$\longrightarrow$	${}^3He_2$ + 1 gamma ray
${}^3He_2 + {}^3He_2$	$\longrightarrow$	${}^4He_2 + 2\,{}^1H_1$ + 1 gamma ray
Net: $4\,{}^1H_1 + 2\,{}^0e_{-1}$	$\longrightarrow$	${}^4He_2$ + 8 gamma rays + 2 neutrinos

Interspersed on the continuous solar spectrum are numerous dark absorption lines produced by heavier elements in the outermost layers and surrounding atmosphere.

If the Sun consumed hydrogen at a steady rate, it would take about 100 billion years for it to exhaust all its fuel. The actual life expectancy of the Sun is a good deal shorter than this because only the hydrogen in the core is hot enough to undergo fusion. It is thought that there is probably enough hydrogen left in the core to keep the Sun burning at more or less the same rate for another 5 billion years. The Sun is therefore now at about the halfway-point in its life.

The Main Sequence

The stars which dot the night sky are actually other suns, but much farther away. They appear only as dimensionless points of light in the telescope. The distances are so great that it is useful to introduce a new scale for space. This is the *light-year,* which is the distance traveled by light in one year. Since the speed of light is 3×10^8 meters per second, the light-year is 9.464×10^{15} meters. The nearest star is Alpha Centauri, 4.3 light-years distant. It is 5000 times further from the Sun than the outermost planet, Pluto.

Generally, the spectrum of the light from a star is a continuous band interspersed with dark lines due to absorption by elements in the outer layers of the star. The details of the spectra differ from star to star. The continuum spectrum of a star is strongly reminiscent of that emitted by an ideal radiator. Different stars have different colors, ranging from deep red, through orange, to yellow, to blue, and finally to white. Some stars even have their maximum emission in the ultraviolet region of the spectrum. The wavelength of light at which the maximum emission occurs establishes the color of the star and hence specifies the *effective temperature.* The cooler stars have a reddish color, whereas the hotter stars are white. In the early twentieth century, scientists at Harvard University developed a scheme by which the spectra of different stars were classified in an orderly sequence. The stars are divided into seven main spectral groups, in order of decreasing temperature. These are denoted by the letters O, B, A, F, G, K, and M. Each major class is divided into ten sub-divisions, designated by the numbers 0 to 9. Class G3 is followed by G4, F9 by G0, etc. The Sun itself is a class G2 star. Generations of astronomy students have memorized the Harvard classification scheme by means of the mnemonic "Oh, Be A Fine Girl, Kiss Me."

A star can also be classified in terms of its intrinsic luminosity. At about the same time that the Harvard classification scheme was

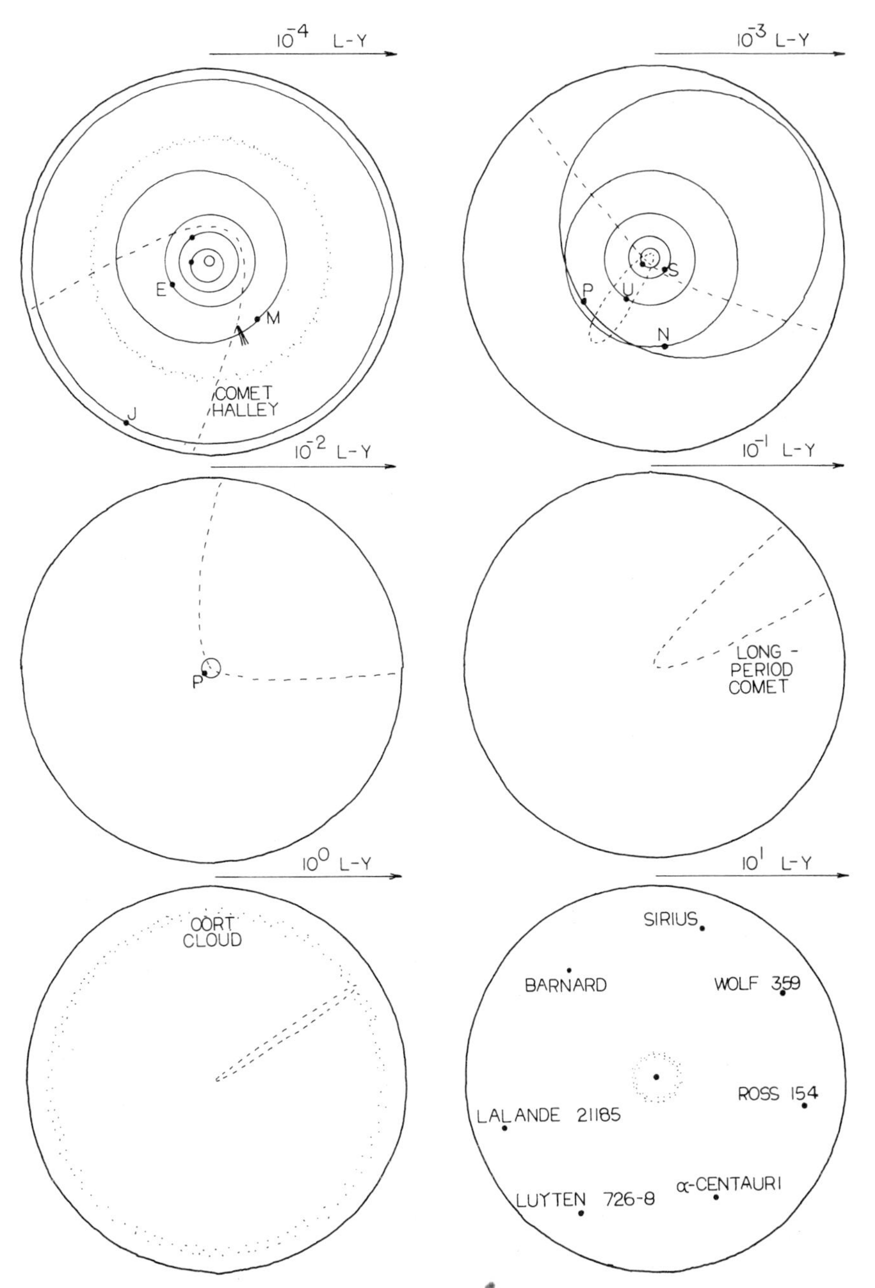

FIGURE 2.2. The scale of space. Each view encompasses exactly 10 times the distance of the previous one. The nearest stars are 5 orders of magnitude farther away than the most distant planet, Pluto. From R. L. Forward (1976). Copyright 1976, British Interplanetary Society.

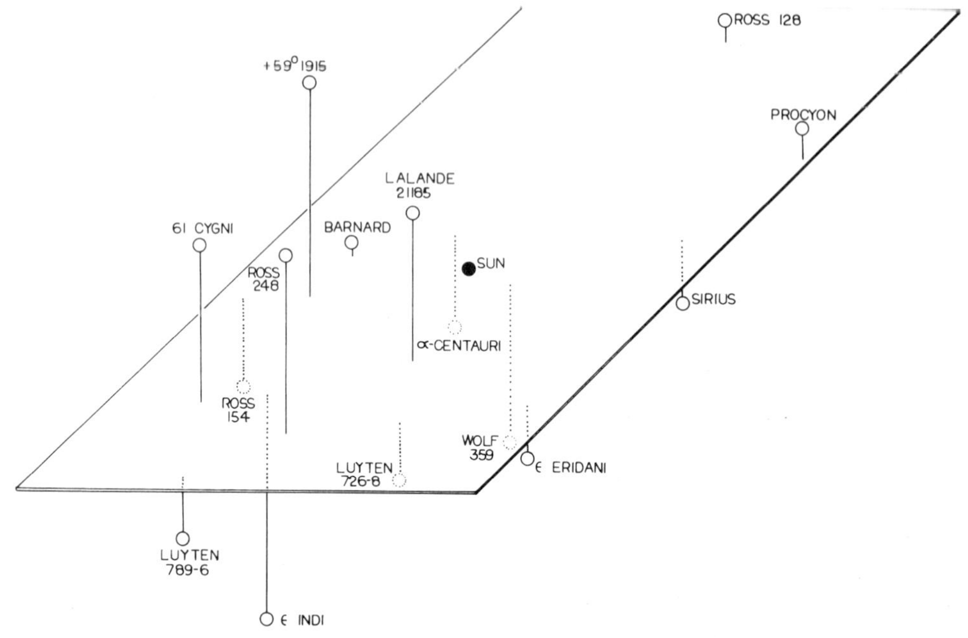

FIGURE 2.3. The location of the fifteen nearest stars in relation to the Sun. The plane shows the orientation of the Earth's orbit around the Sun.

proposed, Einar Hertzsprung of Denmark and Henry Norris Russell of the United States independently demonstrated that there is a strong correlation between the spectral type of a given star and its intrinsic luminosity. This correlation is seen by plotting the spectral type of a particular star on the horizontal axis of a graph and the luminosity of that star on the vertical axis. This type of graph is called a *Hertzsprung-Russell diagram,* and such a plot is shown in Figure 2.4. Note that most of the data points (including the Sun's) fall on a thin strip extending from high-temperature, high-luminosity stars on the left to low-temperature, low-luminosity stars on the right. This narrow band is called the *main sequence.* Full-sky surveys indicate that about 73 percent of all main-sequence stars are of class M, 15 percent are of class K, and only about 12 percent are brighter than class G.

It was at one time believed that all main-sequence stars were "born" on the left-hand side of the H-R diagram and moved to the right along the main sequence band as they aged and cooled. As a result, stars on the left-hand side of the H-R diagram came to be termed "early" and those on the right came to be termed "late." Today it is known that such a picture of stellar evolution is entirely incorrect, but the "early-late" terminology has nevertheless become deeply imbedded in astronomy.

TABLE 2.2 Stars of the Main Sequence

Class	Eff. Temp (C)	Mass (M/Mo)	Luminosity (L/Lo)	Radius (R/Ro)	Density (gm/cc)	Abundance (%)	Life-time (10^9yrs)	Special Details
O	30,000–50,000	>15	11,000–300,000	>6	<0.04	0.00001	0.02–0.001	Bluish white in color. Strong lines from He^+, He^0, C^{2+}, Si^{3+} No H lines.
B	11,700–30,000	2.8–15.0	10–11,000	2.3–6.0	0.25–0.04	0.09	1.0–0.02	Blush white in color. He^+ line vanishes after B5. He^0 line max at B2. H lines more apparent.
A	7700–11,700	1.6–2.8	4.0–10.0	1.4–2.3	0.56–0.25	0.5	2.0–1.0	White in color. Very strong H lines. Lines from heavy metals (Ca^+, Cr^+, Fe^+, Ti^+ Fe^0) increase thruout the class.
F	6200–7700	1.1–1.6	1.0–4.0	1.1–1.4	1.0–0.56	2.9	9.0–2.0	Yellowish white in color. Weaker H lines. Strong Ca^+ lines. Many lines from neutral and singly ionized heavy metals.
G	5200–6200	0.85–1.10	0.4–1.0	0.8–1.1	2.0–1.0	7.3	25–10	Yellow in color. Very strong Ca^+ lines. Many lines from neutral metals. Only very weak H lines. Molecular bands from CH and CN appear.

TABLE 2.2 (Continued)

Class	Eff. Temp (C)	Mass (M/Mo)	Luminosity (L/Lo)	Radius (R/Ro)	Density (gm/cc)	Abundance (%)	Life-time (10^9yrs)	Special Details
K	3200–5200	0.50–0.85	0.07–0.40	0.6–0.8	5.0–2.0	15.1	80–25	Orange in color. Stronger metal lines and more molecular bands. TiO lines appear.
M	2300–3200	<0.5	0.0004–0.07	<0.6	>5.0	73.3	>80	Red in color. Strong neutral metal lines. Stronger TiO lines. Many more molecular bands.

There are a few stars that fall considerably above or below the main sequence. The stars lying below the main sequence are called *dwarfs*. The dim companion of Sirius (Sirius B) is an example. It has a surface temperature similar to that of Sirius A but is nine magnitudes fainter. This must be because Sirius B is much smaller than Sirius A. Dwarf stars are typically only 0.02 to 0.03 solar radii in size, dimensions typical of planets rather than stars. Because of their tiny sizes, dwarf stars are much denser than normal stars. A spoonful of dwarf star "stuff" would weigh over a ton on Earth. Stars like Betelgeuse, Antares, and Capella, lying far above the main sequence, are called *giants*, because they are so very much brighter than main-sequence stars of the same temperature. This is because they are physically much larger. Giant stars range in size from 2.5 solar radii for some of the normal giants and subgiants up to 360 solar radii for some of the supergiants like Betelgeuse. Betelgeuse is so large that if it were substituted for the Sun at the center of the solar system it would engulf the orbits of all the planets out to Mars. Because of their large sizes, giant stars have extremely small densities, often far less than that of air. The giants and the dwarfs are quite rare in comparison with the stars of the main sequence; within 20 light-years of Earth there are no giant stars at all and only nine dwarfs.

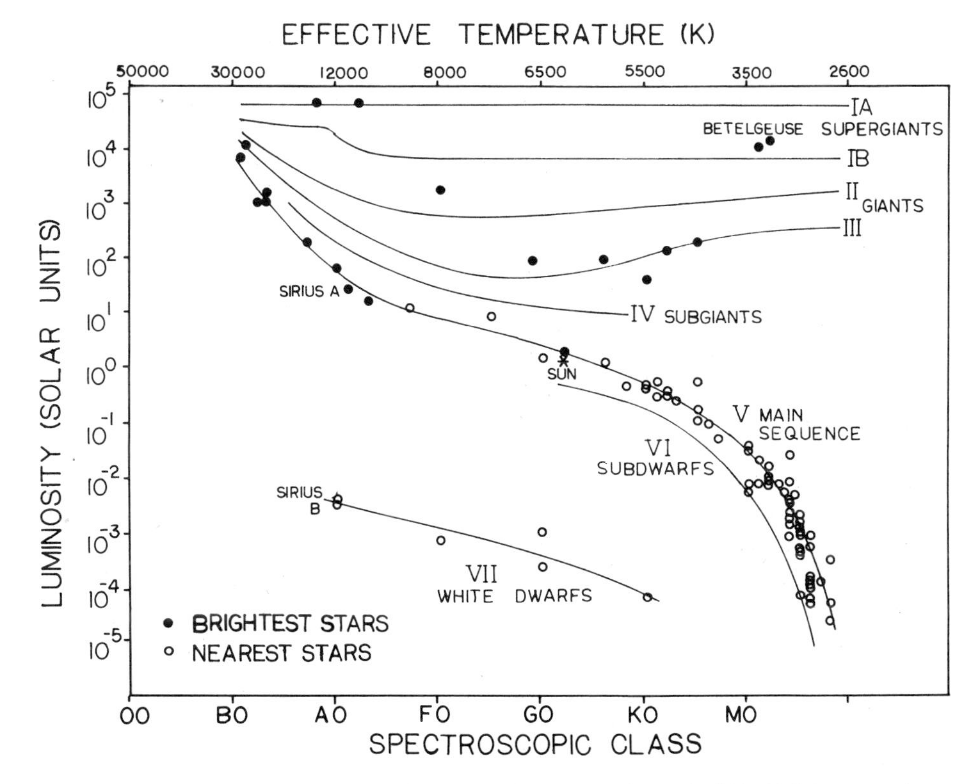

FIGURE 2.4. The Hertzsprung-Russell diagram for the stars located within 20 light-years of the Sun. Also shown are the twenty brightest stars in the sky.

The Birth and Death of Stars: The Forge of the Cosmos

All stars originate in large, massive interstellar clouds made up of hydrogen and helium gas intermixed with slight amounts of ammonia, methane, and water vapor, as well as dust grains rich in metals and silicate minerals. Instabilities are set up by the random motions of other stars, and the clouds develop thousands of vortices, each of which collapses under its own weight to form a star. As a protostar collapses, the interior becomes very hot. Eventually, the temperature becomes so high that the interior hydrogen begins to undergo fusion to produce helium. The protostar becomes self-luminous. At this instant, the main-sequence lifetime of the star begins.

After hydrogen fusion has begun, each star burns at a more or less steady rate, although there is a slow but gradual increase in overall luminosity over many millions of years of time. As the hydrogen in the stellar interior is consumed, helium builds up in the core. A star maintains its overall size and constant luminosity by establishing a delicate balance between the force of gravitational contraction and the production of thermonuclear energy within the interior. So long as there is sufficient thermonuclear fuel within the core, the balance can be maintained.

Once the hydrogen fuel in the core is gone, the delicate balance is upset and the stable main-sequence phase of the star's life cycle comes to an end. The main-sequence lifetime of a star depends on how much thermonuclear fuel is initially present in the interior, as well as on the rate at which it is consumed. Bright, hot stars of classes O and B are 10 to 40 times as massive as the Sun, but they are so luminous that they exhaust their hydrogen fuel within only a few million years. The main-sequence lifetime of the Sun is approximately 10 billion years. Less massive stars of class K and M have such low luminosities that they have appreciably longer main-sequence lifetimes, up to 250 billion years for a small, cool star of class M5.

Once the fuel in a star's core is gone, things begin to happen rather rapidly, the sequence of events depending to a great extent on the original mass of the star. About 4.5 billion years from the present, all the hydrogen in the solar core will finally be exhausted, and the Sun will be left with a helium core at its center. At this time the outer layers of the Sun will begin to expand, and the core will contract and grow hotter. Eventually, the Sun will become 50 to 100 times larger than it is now and will be 300 times more luminous. At this point the Sun will be a red giant. Eventually, the core gets so hot that helium ions can fuse to produce carbon. The Sun's helium-burning phase will be rather brief. Eventually, the helium in the core will become exhausted, leaving a core largely of carbon. During the latter stages of the Sun's death throes, the diffuse outer layer of the bloated red giant Sun will be blown away into cosmic space, enriching the interstellar domain in carbon. When the last of the helium thermonuclear fuel in the dying Sun is gone, there will no longer be any energy to sustain its exterior radius. It then will collapse down to planetary size, forming a dwarf. At this time the remaining Sun will be largely carbon. Over the next few billion years the dead Sun will gradually cool off as it radiates its remaining thermal energy into space until it stops emitting light forever. The Sun will last for about 10 billion years on the main sequence, but will take only a billion years or so to perish after it has exhausted the hydrogen fuel in its core.

When stars more massive than the Sun perish, they can get hot enough in their cores to synthesize successively heavier and heavier elements. If the temperature of the carbon core of a dying star can get hotter than about 500 million degrees Celsius, carbon nuclei can fuse to produce heavier elements, such as oxygen, magnesium, sodium, and neon. If the core can get as hot as a billion degrees Celsius, the oxygen nuclei can react to create even heavier elements, such as sulfur, phosphorus, and silicon. Dying stars larger than 20 solar masses can have core temperatures reaching as high as 2 billion degrees Celsius, allowing silicon to react to form heavy metals, such as nickel, iron, and cobalt. Such massive stars are relatively short-lived on the cosmic scale. They burn fiercely for a brief time, then they die and disperse the heavy elements that they synthesized to the interstellar medium. These newly formed heavy elements gradually drift into large interstellar clouds of gas and dust, ultimately to end up inside new generations of stars which are continually being created out of the dust and debris in space.

The most massive O and B stars have a particularly spectacular death awaiting them. At the end of their lives they literally explode. This type of violent stellar flareup is known as a *supernova.* A supernova rises rapidly to maximum light intensity, fading much more slowly. At the maximum, a supernova can match or even exceed the combined luminosity of all the rest of the stars in the galaxy. When these stars finally use up all of their thermonuclear fuel, a catastrophic collapse takes place. As the outer layers fall inward, the temperature rapidly heats up to billions of degrees Celsius. This rapid temperature rise causes the unburned thermonuclear fuel in the outer layers of the star to ignite, and the star blows up. An intense flood of neutrons, X-rays, gamma rays, and energetic particles is released. Much of the outer material of the star is violently ejected into interstellar space.

After the supernova explosion has ended, a small, incredibly dense object is left behind. This object is known as a *neutron star,* because it consists largely of nuclear matter packed so tightly that densities can exceed a trillion gm/cc, much, much greater even than densities of dwarf stars. Neutron stars typically have radii of only 10 kilometers. The masses range from 1.4 to about 3 solar masses, so a spoonful of neutron star "stuff" would weigh about 10 billion tons on Earth.

At such enormously high densities, matter quite naturally has different properties from ordinary matter as found on Earth. The dense neutron matter may form a superfluid at the interior of the star, with no viscosity and no resistance to flow. The few protons and electrons still left may make this fluid a superconductor as well, with zero electrical resistance and an eternally flowing electrical current. Enormous electrical currents may exist in the interior, creating intense magnetic fields of up to a trillion times as large as Earth's. The outer layers of the star may actually consist of an ultradense, solid crystalline "crust" of free electrons and atomic nuclei. An astronaut foolish enough to land on the surface of a neutron star would find that he weighed 10 billion tons. The enormous gravitational forces would spread both him and his ship out into a thin film covering the entire surface of the star.

Until 1967, neutron stars were entirely theoretical objects, existing only in the pages of esoteric physics journals. However, in that year a group of radio astronomers at Cambridge University in England noted an extraterrestrial radio source pulsing once every 1.337 seconds. The pulses were so regular that they could be used as a clock accurate to one part in a hundred million. For a brief time it was seriously thought that the pulses might have an intelligent origin, perhaps being transmitted as part of some sort of vast interstellar long-range navigation system that guides the starships of alien civilizations back and forth across the galaxy. However, the broad frequency spread (or *bandwidth*) of the pulses seemed to suggest a natural origin, and after much thought it was finally concluded that the pulses originated from a rapidly rotating neutron star. This interpretation is now almost universally accepted. Such an object is known as a *pulsar.* About 200 pulsars are now known, having periods ranging from 0.0015 to 3.0 seconds.

The process of nucleosynthesis within massive stars is believed to cease with iron. All of the thermonuclear reactions which synthesize elements lighter than iron release more energy than that which is required to initiate them. This energy is ultimately given off as the heat and light which make the star shine. However, a net input of energy is required to form still heavier elements, such as copper, mercury, or bismuth, via the fusion of lighter ones. It is believed that these heavier elements were all created during supernova explosions of massive stars. At this time large numbers of highly energetic neutrons are released. The lighter elements in the exploding star capture some of these neutrons and form elements of progressively greater and greater atomic weights. During the explosion many of the heavier elements thus synthesized are ejected into the interstellar medium, ultimately to end up in new stars. All of the heavier metals essential to human technology (e.g., germanium, selenium, copper, barium, palladium, uranium, lead, platinum, gold, and silver) were originally synthesized in supernova explosions many billions of years ago.

Habitable Stars

As we have seen, the stellar population includes all sorts of strange and exotic objects. Which of them are possible homes for living beings? In order for a planet to develop any sort of higher life forms, its central star must have a more or less constant luminosity over periods as long as billions of years. Sudden changes in stellar luminosity, the abrupt appearance of massive flaring, or the emission of strong X-ray or gamma-ray radiation will spell instant death for life on planets surrounding the star. This excludes as possible havens of life such stars as red giants, planetary nebulae, white dwarfs, or neutron stars. In order for a star to be able to support life, it must be a stable member of the main sequence.

It has taken about 5 billion years for intelligent life to appear on Earth, beginning with the time that the Sun first entered the main sequence. The Sun appears to have been shining at about the same rate ever since. Biological evolution on other inhabited worlds should take about the same amount of time. The lifetime of a star decreases as one moves from right to left along the main sequence. Stars of classes earlier than F5 have lifetimes shorter than 5 billion years. They will leave the main sequence and become red giants before their planets have a chance to develop any intelligent life forms. A star such as Altair (class A7) is probably so young that its planets (if there are any) cannot have evolved any life forms more complex than the simplest one-celled organisms. Altair will drift off the main sequence before biological evolution can advance very far.

Stars more massive than F5 can therefore be ruled out in any search for intelligent beings. An entirely different argument has been used to rule out certain low-mass stars, in particular those of class M. These stars are very dim in comparison to the Sun, and in order to sustain any life their planets would have to be quite close. The dimmer the star, the closer the planet must be. For some of the later M-class stars, planets would have to be as close as 0.02 AU in order to be life-supporting. This is probably not impossible, and such miniature solar systems may indeed exist, much like Jupiter and its moons. There is nevertheless a limit. The closer a planet is to its sun, the stronger the tidal force acting on it. These tidal forces will tend to brake the rotation of such a planet until it eventually locks into synchronization with its revolution, keeping the same face forever turned toward its star. Such a planet would roast on one side and freeze on the other. The oceans on the sunward side of the planet would all boil away. On the far side they would all freeze solid. The prospects for life on such planets are exceedingly dim. It is estimated by various workers that stars less massive than about 0.5 solar masses are so faint that any planets close enough to sustain life are locked into synchronous rotation. This restriction rules out all class M stars as sites for life.

Class M stars present an additional problem in that most, if not all, of them are *flare stars*. Such stars can suddenly and unpredictably become up to 4 magnitudes brighter for short periods of time. Such an event would probably be fatal for any life forms on planets orbiting such an unfortunate star. Even if life could actually evolve on planets around class M stars, it would quickly be snuffed out by a massive flare.

Stars more massive than F5 and less massive than M0 can therefore be eliminated in any search for extraterrestrial civilizations. Only about 2 percent of all stars are more massive than F5, so this restriction does not reduce the number of possible life-supporting stars very much. However, 73 percent of all stars in the Solar neighborhood are of class M, and the elimination of these does restrict the number of potential sites very severely. This leaves only 25 percent of all stars capable of ever giving rise to intelligent life forms. Stars of spectral types between F5 and M0 are said to be *sunlike*. The fraction (designated by f_{sun}) of stars that are sunlike is therefore equal to:

$$f_{sun} = 0.25$$

THE RIGHT POPULATION

If you go outside on a clear night (well away from city lights), you can see a dim band of light stretching all across the sky. It is called the *Milky Way* because of its obvious appearance to the eye. It is not uniform in appearance; there are very bright patches as well as dark spots. When Galileo turned his telescope toward the Milky Way, he discovered that the light patches were actually composed of many thousands of stars which could not be separately resolved by the naked eye. Since the Milky Way completely surrounds the sky and divides the celestial sphere into two hemispheres, the Sun itself must be imbedded within this vast stellar aggregate.

The structure of the galaxy as presently understood is shown in Figures 2.5 and 2.6. It is a giant flattened disk of about 200 billion stars. The radius of the disk is about 80,000 light-years, with the Sun about 28,000 light-years from the center. The stellar disk is about 3000 light-years thick. Near the equator of the disk is a 500-light-year-

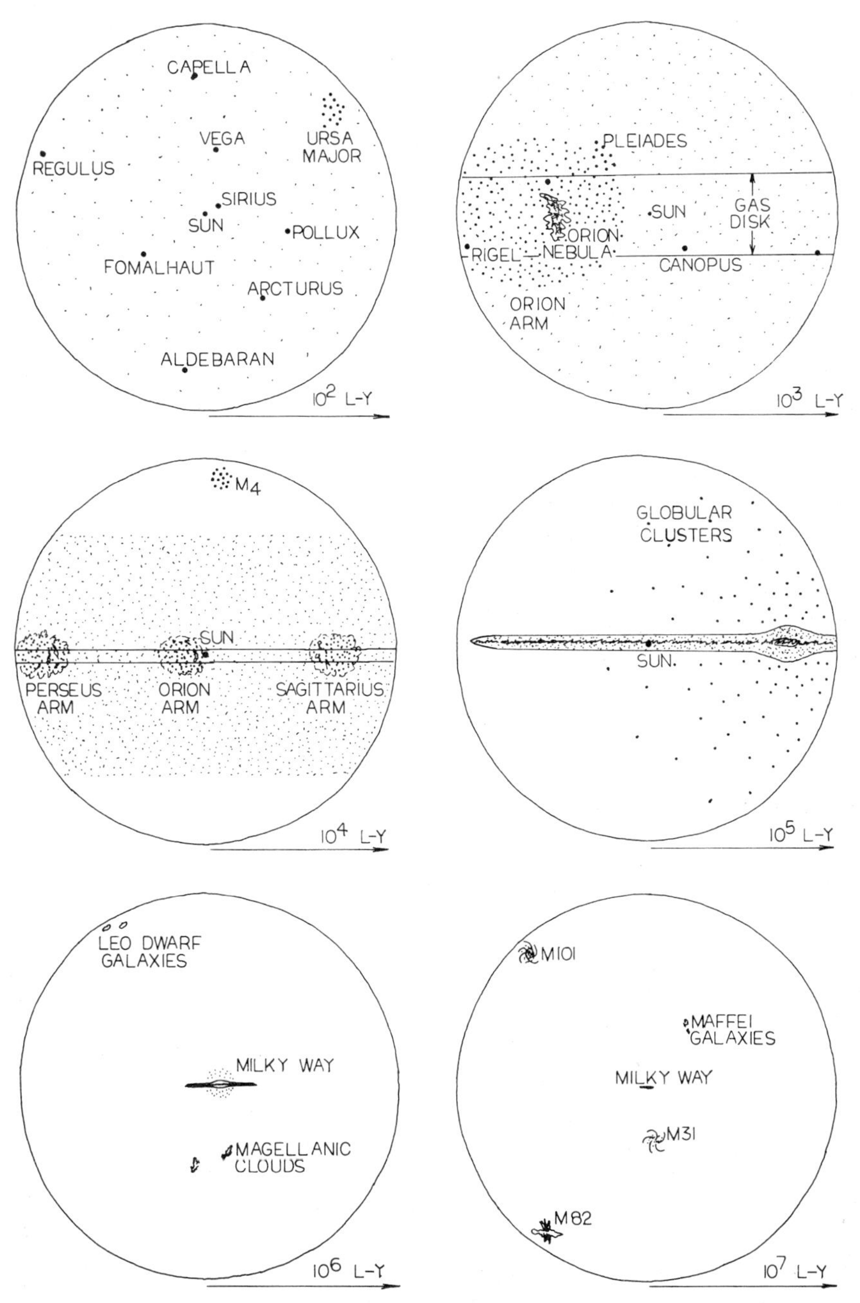

FIGURE 2.5. The scale of interstellar space. Each view encompasses exactly ten times the distance of the previous one.

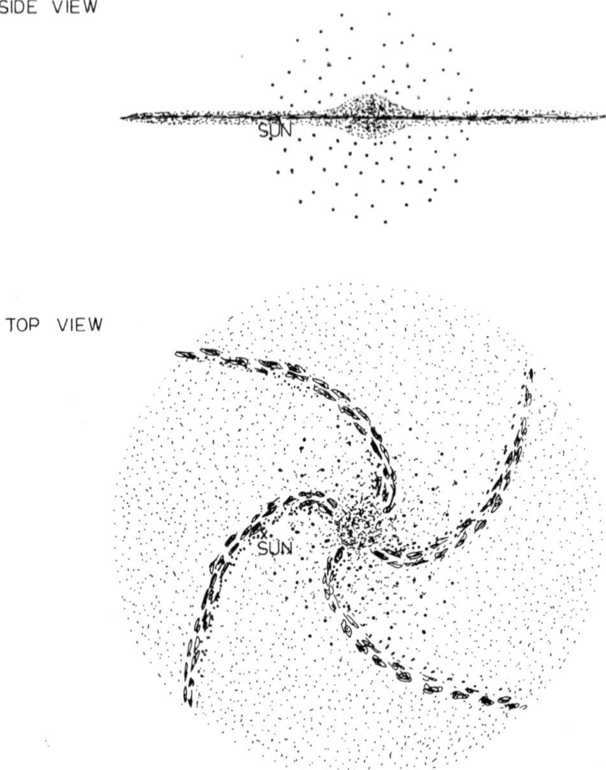

FIGURE 2.6. Horizontal and vertical views of the Milky Way Galaxy.

thick layer of hydrogen gas within which our own solar system is imbedded. It is estimated that light from stars more than 10,000 light-years distant in the galactic plane is prevented from reaching the Earth by this layer of gas, although visibility is much better outside the galactic plane. Much of the galaxy is thereby hidden from our view, making the study of its structure quite difficult.

The stars nearest the Earth appear to be distributed randomly in space, but at larger distances significant nonuniformities are present. At distances of 100 light-years or more, loosely bound, *open clusters* of stars are found. These clusters congregate near the galactic plane and have a mean separation of about 300 light-years. They range in size from 3 to 30 light-years across. The best-known open cluster is the Pleiades, about 400 light-years distant. Open clusters have lots of hot, bright stars of classes O and B. There is usually also a lot of loose gas and dust around these stars. Open clusters are relatively young objects on the cosmic scale and appear to be the cradles of newly born stars which only recently have condensed out of vast interstellar clouds of gas and dust. The member stars are relatively young, some only a few million or even a few thousand years old. As time passes the stars in an open cluster will individually drift away, so that in a few million years the cluster will totally disperse.

At far greater distances a completely different type of star cluster can be found. This is the *globular cluster,* the nearest of which is 9000 light-years distant. They are much denser than open clusters and have far larger numbers of stars. Their mean diameter is about 30 light-years, but they are estimated to have many millions of stars. The stellar density at the center is typically as high as 1000 times that in the solar neighborhood, which means that stars located there could be as close as 0.5 light-year apart. Unlike open clusters, there is almost no gas or dust in globular clusters. They have no hot, bright stars of classes O or B, but they do have a lot of red giant stars. The spectra of the light emitted from stars within globular clusters show that they generally contain much smaller concentrations of heavy elements than do the stars in the immediate solar neighborhood. The stars in globular clusters must be quite old, having formed very early in the history of the galaxy when the heavier metals were in far less abundant supply.

Unlike open clusters, globular clusters are rarely found near the galactic plane. The 120 or so globular clusters that can be seen are distributed in a nearly spherically symmetric array of maximum radius 60,000 light-years. The center of the spherical array of globular clusters seems to be coincident with the geometrical center of the main galactic disk. The metal-poor globular clusters tend to congregate in the outermost parts of the spherically symmetric array, with clusters that are richer in metals being closer to the galactic center. They all circle around the galactic center in elliptical orbits, and each cluster passes through the plane of the galactic disk about once every billion years or so. There is also a diffuse "halo" of unclustered stars in this general region, probably produced by individual stars that managed to escape from globular clusters many millions of years ago. They have up to 100 times less carbon within their interiors than does the Sun.

The entire galaxy is undergoing a slow rotation. The velocity with which other parts of the galaxy are moving with respect to Earth can be measured in some detail by examining the Doppler shifts in the 21-centimeter microwave emission from the hydrogen gas along various lines of sight. Since the galaxy is not a rigid body, different parts are moving at different rates. Every star, gas molecule, and dust particle in the galaxy travels in an independent Keplerian orbit about the center. The Sun itself travels in a nearly circular orbit at an average speed of 220 kilometers per second, taking about 225 million years to make a single circuit. The Sun's orbit is nearly coincident with the plane of the galactic disk.

The outer regions of the galactic disk are split into at least four spiral arms. Individual arms are about 1500 light-years thick and are 60 to 80 thousand light-years long. Within these arms, the densities of stars and dust are a good deal higher than outside. Very young, hot stars are abundant in these arms. The inner margins of the galactic spiral arms are marked by dark, gaseous nebulae which perhaps signal the onset of the compression of interstellar gas by a pressure wave that advances radially outward throughout the galactic disk. All new star creation is believed to take place inside these dense clouds within the spiral arms. After the stars are born, they individually drift out of the arm in which they were formed and pass into another. The Sun is near the inner edge of one of these arms, about 600 light-years from its center. This arm is sometimes called the *Orion Arm,* as the Great Nebula of Orion is a prominent member.

The galactic center is located in the constellation Sagittarius, but is hidden from our view by gas and dust. It can nevertheless still be studied by astronomers, since it happens to be the brightest radio source in the sky. At the very center of the galactic disk there is a slightly flattened bulge of stars about 16,000 light-years thick. The bulge consists of a dense cluster of old stars imbedded within a thin matrix of gas and dust. The stellar density steadily increases as one gets closer to the galactic center. Although most of the stars in this central bulge are old, there is a steady increase in the number of new, metal-rich stars as one nears the galactic center.

The central few light-years of the galaxy (termed the nucleus) may actually contain many millions of stars packed as close as 0.01 light-year apart. Stellar collisions could be taking place at average intervals as short as a thousand years. There seems to be something rather odd going on at the very center of the nucleus. Some sort of supermassive object seems to be present there, one perhaps as large as 50 million solar masses. This central object seems to be coincident with a highly compact intense radio source known as Sagittarius A, which is only 0.001 arc second in angular diameter. The radio emission from the nucleus varies irregularly in intensity from one day to the next, but changes very little on time scales of minutes. As judged by the time scale of the intensity variations, the radio source can be no larger than a light-day across. This is approximately 100 AU, not much bigger than our solar system. The galactic nucleus is also an intense source of highly energetic X-ray and gamma radiation, which appears to be coincident with the compact radio source. The total power being emitted from the galactic nucleus is estimated to be about thirty million times that which is emitted by the Sun. A source more energetic than a million suns is packed into a space no larger than a solar system! Theoretical astronomers propose that the only object that could conceivably produce such intense energy emission from so small a space is a massive black hole.

The galaxy is estimated to be from 12 to 14 billion years old, much older than the Sun. The galactic precursor was probably a nearly spherical cloud of hydrogen and helium gas of at least a trillion solar masses. The protogalactic gas cloud was undergoing a slow rotation, and under mutual gravitational attraction it slowly began to collapse. The first stars began to condense almost immediately thereafter, forming the globular clusters and the halo stars, as well as the oldest stars found in the galactic center. These initial stars formed a more or less spherically symmetric array about the center of the collapsing cloud, and because of their early birth they contained very few elements heavier than helium. Such stars are said to be of *Population II*. As the more massive Population II stars exploded and dispersed their materials into outer space, the interstellar medium was steadily enriched in heavy metals as the galactic cloud continued to contract. Because of viscous drag, the gas and dust that had not initially condensed to form the early Population II stars gradually drifted farther inward and eventually settled down into a flattened disk perpendicular to the galactic rotation axis. Later generations of stars subsequently formed within this flattened disk. These stars are termed *Population I* and are much richer in heavier elements. The division of the galaxy into spherically symmetric Population II stars and cylindrically symmetric Population I stars is shown in Figure 2.6. About two-thirds of all stars in the galaxy are Population II, with one-third Population I.

Which of the two populations of stars would be the better prospect for the presence of intelligent life? Even if Population II stars have planets, these worlds must be almost totally devoid of metals or silicates. Any planets present would presumably be like Jupiter, consisting almost entirely of hydrogen and helium. Planets like Earth, having metallic cores and silicate crusts, would be quite rare. The hydrogen-rich but metal-poor planets in globular clusters would make poor prospects for the presence of any type of life at all, to say nothing of intelligent technological civilizations capable of space flight or interstellar communication. For all we know, incredibly ancient and wise civilizations might be found on planets around stars in globular clusters, but it is difficult to imagine how their inhabitants could ever develop an advanced technology without a supply of metals. Since the nearest globular cluster is 9000 light-years away, the prospect of any sort of contact with civilizations located there is exceedingly remote.

Attention can therefore be safely confined to Population I stars, which are largely concentrated within the outer parts of the galactic disk. Population I stars are second- or even third-generation objects, having formed from the debris created by the explosions of extremely massive, short-lived stars that enriched the interstellar medium in heavy elements when they died. Planets around such stars should have an adequate supply of nitrogen, oxygen, and carbon from which to create living beings, as well as an abundant amount of heavier metallic elements from which a sophisticated technological society could be built. The fraction of the stars in the galaxy that are members of Population I is approximately a third. So there about 70 billion stars in the galaxy that are young enough to have planetary systems with a sufficient supply of heavy elements to permit the emergence of life and the subsequent establishement of an advanced technology. The fraction of stars in the galaxy which are of the correct population is:

$$f_I = 1/3$$

PLANETS AROUND OTHER STARS

Do other stars have planets? The stars are so far away that it has so far proved impossible to observe any objects as small as planets orbiting them. Therefore it cannot yet be conclusively demonstrated that any star other than the Sun has planets. Nevertheless, there is a good deal of very strong *indirect* evidence that other stars do possess planetary systems and that planetary formation is a natural consequence of stellar evolution.

The most controversial evidence for the existence of extrasolar planetary systems has been that which has been gathered from astrometric measurements of the motions of other stars. In 1963, Peter Van de Kamp, the director of the Sproul Observatory at Swarthmore College, announced that he and his colleagues had discovered that Barnard's Star (a faint star 5.9 light-years from Earth) has an invisible companion that was of sufficiently low mass to be considered a planet rather than a "black dwarf" subluminous star. Van de Kamp had analyzed 2413 photographic plates taken of this star since 1938 and had noted that there was a small periodic oscillation in the position of the star in the sky. Detailed analysis of the oscillations indicated that the suspected planet had a mass of about 1.5 Jovian masses and was in an eccentric orbit averaging 4.4 AU from the star. Van de Kamp later refined his analysis and concluded that the data could be better understood if there were actually *two* Jovian-sized

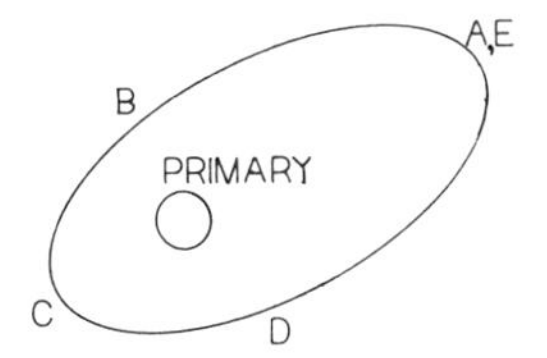

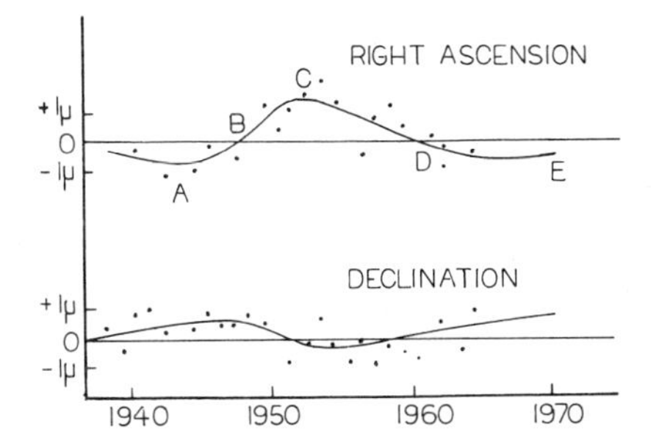

FIGURE 2.7. Van de Kamp's graph of the wobbling motion of Barnard's Star. This may be evidence for the presence of a Jovian-sized planet around this star. From A. T. Lawton (1970). Copyright 1970, British Interplanetary Society.

planets orbiting the star in circular, coplanar orbits. The initial announcement caused a sensation, and many other astronomers began to pay closer attention to this small star. Some astronomers imagined that they could see evidence of as many as five planets orbiting Barnard's Star. Other workers could not observe the oscillations at all, concluding that the purported "wiggles" of Barnard's Star were nothing more than systematic errors made over the years by the Sproul Observatory refracting telescope. The errors involved in making accurate measurements of the position of a star over long periods of time are large and are of the same order as the reported magnitudes of the oscillations themselves. As a result, the proposed planetary system of Barnard's Star has become a subject of hot controversy within the astronomical community, with the majority opinion tending to dismiss the observation as an artifact.

Since the initial announcement of the discovery of a possible planet around Barnard's Star, many other stars in the immediate solar neighborhood have been examined for astrometric evidence of the presence of planetary systems. At least 24 of them have been reported to show "wiggling" motions that may indicate that planets are present. However, some, or indeed all, of these measurements may very well be spurious. Astrometric studies of the motions of stars are very difficult to make, and unambiguous detections of oscillations caused by unseen companions have usually been possible only when the mass of the companion is relatively large (of the order of 0.01 to 0.1 solar masses or greater). The wobbling motion of the Sun caused by the motion of the massive planet Jupiter is only about 0.017 seconds of arc as seen by a hypothetical observer near Alpha Centauri. This motion could not be detected by any Earthbound telescope currently in operation.

In seeking evidence of extrasolar planets it might be profitable to look for stars which may presently be forming planetary systems, just as the Sun did nearly 5 billion years ago. Such a search would most fruitfully be carried out in open star clusters or gaseous nebulae, where there are many young, hot stars as well as a lot of stars surrounded by gaseous and dusty envelopes. Light scattering from the dust particles will produce an "infrared excess" in the spectrum of the star, in which there is a good deal more infrared light than if the dust were not there. The irregular brightness of some of these stars may be caused by dust clouds in various stages of collapse that surround them. Many of the fainter members of open star clusters fall slightly below the main sequence, which may mean that they are surrounded by shells of dust. There is some evidence that these dust shells tend to collapse into thin disks very early in the lives of the parent stars. There are even a few observations of what may have been a sudden clearing of gas and dust in these disks, as in the case of the stars FU Orionis and V1057 Cygni. In 1977, astronomers at the University of Arizona discovered a forming star (designated by MWC 349) in the constellation Cygnus that appears to be surrounded by a flat disk that may be a planetary system in the process of formation. A star known as RU Lupi was noted by a group of Swedish astronomers to be rising and falling in brightness at irregular intervals every few days, apparently because clouds of obscuring material were passing across the face of the star. The supergiant star Epsilon Aurigae is eclipsed once every 27 years by an invisible companion that cannot be a spherical body but must instead be a flattened disk that may be the precursor of a planetary system. In 1983, the infrared astronomy satellite IRAS (a joint effort of Great Britain, Holland, and the United States) found that excessive amounts of infrared light were coming from the young bright stars Vega and Fomalhaut. This infrared excess appears to come from a "cocoon" of rocks and dust particles that surrounds these stars. It is thought that these shells are made up of particles larger than a few millimeters in size, perhaps ranging in size from grains of sand all the way to asteroid-sized chuncks of rock. It is possible that the dusk clouds are planetary systems currently in the process of formation. It seems that the formation of gaseous and dusty accretion disks is a natural consequence of the evolution of very young stars. It is possible (although not proven) that many of these accretion disks naturally evolve into planetary systems.

There is thus strong if indirect evidence that many if not all stars produce planetary systems as a natural consequence of their evolution. The fraction of all Population I stars in the galaxy (designated by f_p) that have planetary systems can be assigned the value of unity:

$$f_p = 1$$

MULTIPLE STAR SYSTEMS

About half of all stars within 20 light-years of Earth are actually members of *binary* or *multiple* systems, so such stars must be quite important in the overall pattern of stellar evolution and development. There are basically four different categories of multiple stars, classified according to how they are detected and observed. These are *visual, astrometric, spectroscopic,* and *eclipsing.*

Visual multiple star systems are those in which individual members are far enough apart that they can be seen as separate and distinct points of light in the telescope. If a visual binary such as Kruger 60 is watched for a long time, it can be seen that each of the two individual stars orbits the common center of mass of the system, much in the same way that the Earth orbits the Sun. Unlike the Sun and the planets, however, the masses of the two component stars in a binary system are usually sufficiently similar so that both stars appear to undergo periodic motion in the sky, each traversing a separate elliptical orbit with the center of mass located at the common focus.

The multiple system closest to Earth is Alpha Centauri. It appears to consist of three stars, with components A and B orbiting about each other with a period of 80 years and a mean separation of 23.5 AU. The third member, Proxima Centauri, is 10,000 AU distant

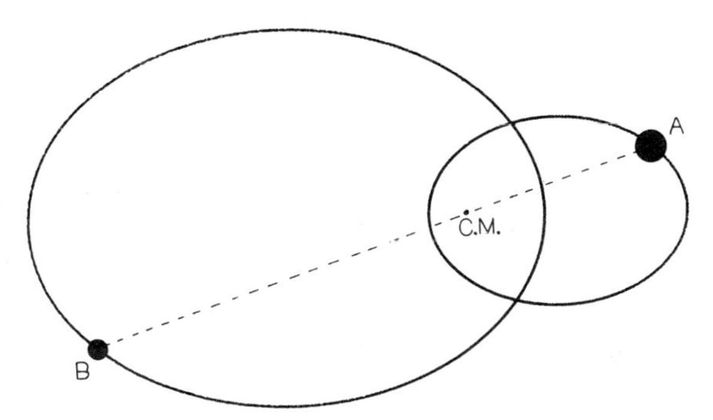

FIGURE 2.8. Orbital motions of the two components of a binary star system. Each component travels in an elliptical orbit with the focus at the center of mass of the system.

from the other pair, but its orbit has not been well determined. It may not actually be physically bound to the other two. Of the 100 stars within 20 light-years of Earth, at least 32 are situated in visual binary systems, and six are in triple systems. Those nearby binary stars whose orbits can be studied in any detail tend to have an average component separation of about 20 AU. Unlike planetary orbits, the orbits of the members of visual binary systems tend to be rather eccentric.

Sometimes only one component of a binary star system can actually be observed in the telescope, the other being too faint to be seen. The presence of the unseen companion can nonetheless be inferred by watching the motion of the visible star. If the invisible companion is sufficiently massive, the visible component will appear to wobble periodically back and forth in the sky in an elliptical path. By studying this motion, the mass of the unseen companion can be measured and its orbit determined without its ever being seen in the telescope. Such systems are referred to as astrometric binaries. In the nineteenth century, the German astronomer Friedrich Bessel noted that Sirius and Procyon, two particularly bright nearby stars, both had periodic motions that could be explained by the presence of massive unseen companions orbiting quite close by. Both companions were visually confirmed later in the century as improved types of telescopes became available.

When the two components of a binary star system are sufficiently close together, no telescope can separately resolve them. Such a system can still be identified as being double by studying the spectrum of the light emitted. Each of the stars has a characteristic spectrum with specific absorption lines. As the two stars rotate about their common center of mass, at any one time one of the pair will be moving toward the Earth and the other away. The wavelengths of the absorption lines from the two stars will therefore be Doppler-shifted by different amounts. The absorption lines from the two stars will appear to oscillate back and forth across the spectrum with the same period as the orbital motion, the lines from each star moving in opposite directions. Double stars identified by observing the periodic motion of their spectral lines are referred to as spectroscopic binaries. About 700 spectroscopic binaries are now known, 3 of which are within 20 light-years of Earth. In order to be detected in this manner, the two components of a spectroscopic binary must be revolving about each other quite rapidly. This means that such binary stars are typically less than 0.1 AU apart, much closer than the members of a visual binary system. Their orbital periods are typically measured in days (or even hours) rather than in years. Of the 100 brightest stars, about 25 are known to be spectroscopic binaries. It is estimated that about 4 to 7 percent of all class F, G, and K stars are spectroscopic binaries.

The final type of binary star is the eclipsing variety. It is a double system in which the plane of the orbit happens to be very nearly parallel to the line of sight from Earth, so that the two components periodically pass in front of each other. The best-known example is the star Algol (Beta Persei), located 88 light-years away. Algol becomes briefly dimmer for a few hours once every three days. In 1782, the English astronomer John Goodricke suggested that Algol was a closely spaced double star system, with the cause of the periodic fading being the passage of a fainter component in front of a brighter one. This model was confirmed in 1889 when the German astronomer Hermann Vogel found that Algol was also a spectroscopic binary, with the spectral lines moving to and fro with the same period as the light variation. More refined calculations have established that the Algol system is inclined at an angle of about 9 degrees with the line of sight, so the eclipse is partial rather than total. The brighter component is a class B8 star (3.4 solar masses), and the dimmer one is class K0 (0.73 solar mass). The orbits of the two stars are nearly circular, and they remain a constant distance of 0.07 AU apart. About 4000 eclipsing binaries of the Algol variety have been catalogued. The individual components of eclipsing binaries tend to be closer than 0.1 AU; the nearer the pair, the more likely an eclipse will be visible from Earth.

Can life-supporting planets be present in binary or multiple star systems? A planet in a binary system will receive light from both components as it travels along its orbit, and the light intensity at its surface may vary to such an extent that it is too hot at one time and too cold at another. Another possible problem is that planetary orbits within binary star systems may be highly unstable. Long before any sort of advanced life can evolve on a given planet in a binary star system, the complex motions of the individual stars may perturb the planet's orbit into so eccentric an elliptical path that it either eventually crashes into one or the other of the stars or else is ejected from the star system altogether. Either eventuality would be fatal for life.

Robert Harrington, of the U.S. Naval Observatory in Washington, D.C., suggests that there might be two exceptions to this rule. By the use of computer simulations of planetary orbits within double-star systems, he found that if the two stars were sufficiently far apart (greater than 20 AU), a stable planetary orbit of radius 1 AU about either one of the components is possible. In this case the other component is sufficiently distant so that its light is much too feeble to have any significant influence on the planet's surface environment. If the Sun had an identical companion at the distance of Pluto, it

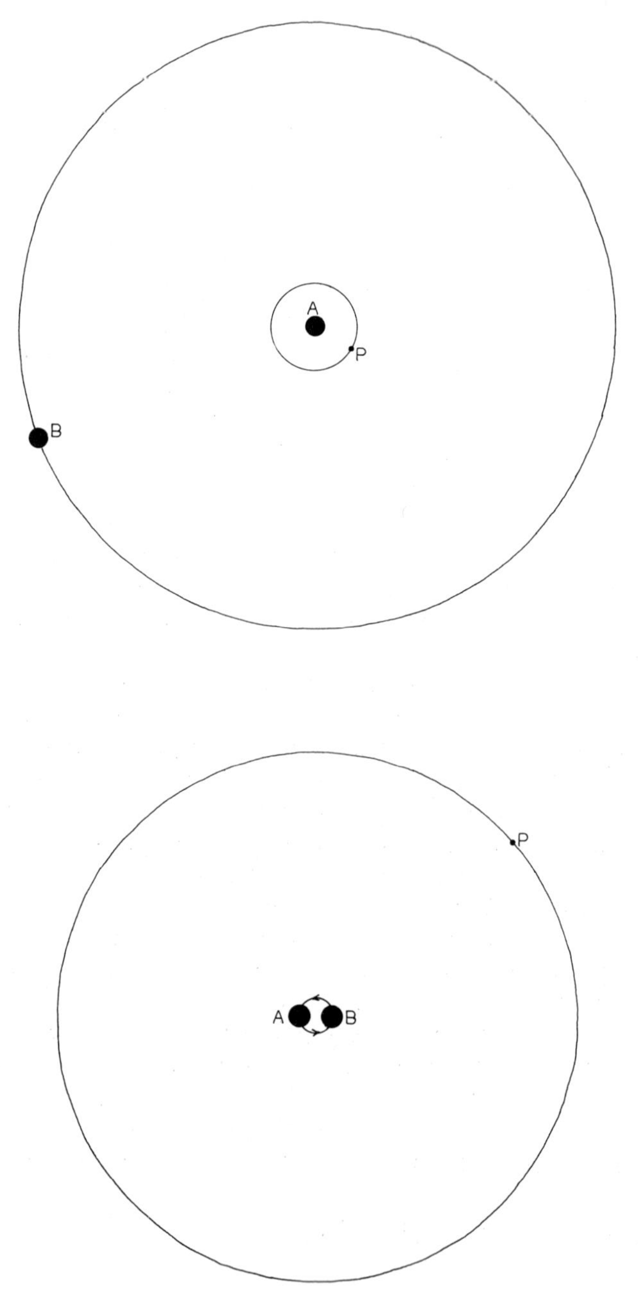

FIGURE 2.9. Stable planetary orbits within a binary star system. Planets may be able to orbit about one star if the other component is sufficiently distant. Planets may be able to orbit about both components if the stars are sufficiently close to each other.

probably would not perturb Earth's orbit to any significant extent nor affect Earth's weather very much. Harrington found that stable planetary orbits in binary systems are also possible if the two stars are so close together that a planet can orbit around both simultaneously. If the Sun were actually a close binary with components less than 0.2 AU apart, life on Earth could be quite tolerable, as the light intensity should not vary significantly as the two stars rotated about each other.

Some writers have suggested that binary or multiple star systems can be ruled out as possible havens of life because they probably do not have planetary systems in the first place. They argue that the formation of a binary star system makes it impossible for a gaseous and dusty accretion disk to remain stable long enough for planets to condense. As of yet there is no firm evidence of any planetary system other than our own, so it is impossible to verify this view. However, the invisible companion of Epsilon Aurigae appears to be a protostar with a surrounding accretion disk which may be the precursor of a planetary system. Planetary systems may be able to form around individual stars in multiple star systems if the stellar components are far enough apart so that their tidal forces do not disrupt accretion disks before they have a chance to condense into planets. Most binary star systems seem to meet this condition. Furthermore, binary stars tend to have highly eccentric orbits whereas planetary orbits (at least in this solar system) tend to be nearly circular, and it may be true that the mechanisms of binary star and planetary formation are so completely different that the existence of one need not necessarily rule out the other. In our own solar system Jupiter and its moons can be considered a miniature solar system, condensed from an accretion disk gathered around a body that could even have formed a small star had it been 10 to 50 times more massive. Obviously the presence of the Sun only 5 AU away did not prevent this from happening.

THE ECOSPHERES OF SUN-LIKE STARS

The Earth receives about 1.4 kilowatts of steady solar power for every square meter of surface area when the Sun is directly overhead. Part of this light is reflected back into outer space, and the rest is absorbed. There is a delicate balance between absorption and reflection, so that the temperature of the Earth has remained essentially constant for billions of years. The geological record indicates that there has been liquid water flowing on the surface for at least 3.8

billion years, which must mean that the average surface temperature has remained between 0 and 100 degrees Celsius all throughout this period. Were the temperature suddenly to rise by an appreciable amount, all of the water in the oceans would evaporate. Were it suddenly to get a good deal colder, the water would all freeze solid to the surface. Either eventuality would be fatal to life. The ability of the Earth to have kept its temperature so well regulated over such a long period of time is quite remarkable and is perhaps the single most important factor that has permitted life to evolve and flourish here on this planet.

The Right Distance

The Earth's present location vis-à-vis the Sun is undoubtedly a statistical quirk of its initial formation nearly 5 billion years ago. A slightly different set of circumstances could conceivably have resulted in the Earth having formed at a different distance from the Sun. How much farther from the Sun could the Earth have originally been placed before the sunlight would have been so dim that the oceans would all have frozen? How much closer to the Sun could the Earth have formed before the oceans would all have boiled away? The range of Earth–Sun distances over which life-sustaining conditions

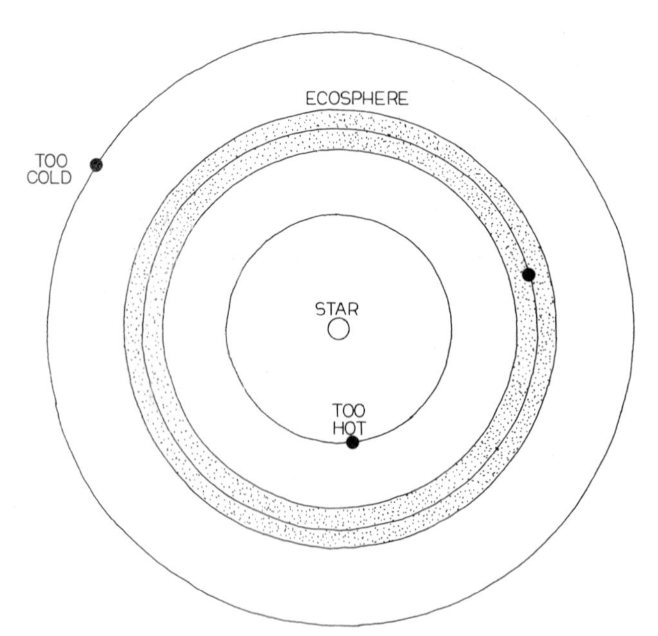

FIGURE 2.10. The stellar ecosphere, the region about a star in which a planet must orbit in order to be habitable.

could still be continuously maintained is called the *ecosphere* of the Sun, a term originally coined by Stephen Dole of the Rand Corporation. The size of the Sun's ecosphere will depend on the intrinsic solar luminosity as well as on the overall response of the Earth to different temperature regimes.

Since very little is known about how the Earth's surface would have evolved had its average temperature been significantly different, the exact size of the Sun's ecosphere is a matter of controversy. We can get a rough idea by looking at the planets in our own solar system. Venus is too close to the Sun; the surface was so hot that its water boiled away into the atmosphere and was subsequently lost to space. Mars is too distant; water at its surface is present only as solid ice. The inner boundary of the solar ecosphere should be placed somewhere between the orbits of Venus and Earth, whereas the outer boundary must be between the orbits of Earth and Mars, but their precise locations are somewhat a matter of conjecture. If one adopts the probably overly optimistic view that the Sun's ecosphere includes the entire space between Venus and Mars, then the inner and outer boundaries would be at 0.75 AU and 1.5 AU respectively. Over this width the intensity of sunlight varies from 1.7 times that on Earth at the inner edge to only 0.45 times Earth level at the outer. Stephen Dole estimates that the Sun's ecosphere ranges from 0.86 AU to 1.24 AU, corresponding to a variation in light intensity of 0.65 to 1.35 times Earth level. Several other rather similar estimates of the width of the Sun's ecosphere have been made. Unfortunately, all of them suffer from the defect of being based on very little hard evidence, depending more on the biases and prejudices of the individual workers (in particular whether they accept or reject the possibility of extraterrestrial life) than on verifiable scientific fact.

The Runaway Climate

Recently, computer simulations of the evolution of Earth's atmosphere by Michael Hart of the National Aeronautics and Space Administration (NASA) have shown that the solar ecosphere may be much narrower than previously thought. The computer model took account of the slow increase in the solar luminosity, changes in the Earth's reflectivity, variations in the amount of living matter present, and changes in the atmospheric content, as well as of a variety of geochemical processes. Hart assumed that the Earth originally had a highly reducing atmosphere of methane, ammonia, water vapor, carbon dioxide, hydrogen sulfide, and nitrogen. Over the first billion years of Earth's existence, oxygen gas released by the ultraviolet photolysis of water vapor high in the atmosphere reacted with ammonia and methane to produce nitrogen and carbon dioxide. The atmosphere was steadily depleted of its hydrogen-rich gases and gradually became less reducing. The carbon dioxide in the air was steadily depleted by reacting with the metals dissolved in the water, forming such minerals as limestone in sedimentary deposits. All throughout this period the surface temperature was a good deal warmer than at present because of the greenhouse effect produced by the heat-trapping properties of the ammonia, water vapor, and carbon dioxide still present in the air in appreciable quantities. Had the Earth been only a few percent closer to the Sun (less than 0.96 AU), a "runaway greenhouse" effect would have taken place about 4 billion years ago. The Earth would have ended up much like Venus, with a thick, dense atmosphere of carbon dioxide and superheated steam. There would be no liquid water on the surface. Under such circumstances, life could never have appeared on Earth.

Fortunately for us, the Earth was far enough from the Sun so that such a catastrophe did not occur. About 2 billion years ago the last of the ammonia and methane that had been present in the atmosphere during the early years of Earth's history were finally cleared away. At approximately the same time, oxygen-generating photosynthesis appeared. The oxygen content of the air began to rise sharply, and the carbon dioxide content decreased more rapidly. The sudden appearance of an oxidizing atmosphere at this time saved the Earth from Venus's fate, and the temperature of the Earth then began to fall rapidly. However, at this point another even greater danger lurked. The sudden drop in temperature caused solid ice to form on the surface, and the Earth acquired its first polar ice caps. Glaciers appeared for the first time. Had the Earth been only 1 percent farther from the Sun (more than 1.01 AU), there would have been a runaway glaciation at this time. The lower temperatures would have produced more surface ice, which would have reflected a greater fraction of the solar energy back into space, resulting in still lower temperatures and even more ice. The oceans would have frozen over, killing all the single-celled organisms then living in the Precambrian seas. The Earth would have ended up like Mars, with all its water in the form of solid ice.

The size of the Sun's ecosphere seems to have been grossly overestimated by earlier workers, who neglected the possibility of thermal runaways and ignored the gradually increasing solar luminosity. According to Hart's model, the Sun's ecosphere is only about 0.06 AU wide. The presence of life on Earth today is the happy result of the rare coincidence of the Earth being located at precisely the right distance from the Sun. Had the Earth formed just a few percent either closer or farther from the Sun, we would not be here.

TABLE 2.3 Estimate of Width of Ecosphere by Hart Model of Atmospheric Evolution

Star Class	Mass (solar units)	Luminosity (solar units)	Outer Boundary (AU)	Inner Boundary (AU)	Width of Ecosphere (AU)	Probability of Planet in Ecosphere
F5	1.24	2.40	1.79	1.68	0.11	0.28
G0	1.10	1.50	1.15	1.08	0.07	0.18
G2(Sun)	1.00	1.00	1.01	0.96	0.05	0.13
G5	0.90	0.70	0.87	0.84	0.03	0.08
K0	0.85	0.50	0.629	0.628	0.001	0.003
K5	0.63	0.20	0.42	0.42	0.000	0.000
M0	0.50	0.06	0.27	0.27	0.000	0.000

The narrowness of the Sun's ecosphere has important implications for the possibility of life-supporting planets around other sunlike stars. The narrower the ecosphere, the less likely that a planet will be located inside it. Michael Hart has performed similar computer analyses for planets identical to Earth situated around other sunlike stars, and the results are shown in Table 2.3. Hart's stellar ecospheres are more than an order of magnitude narrower than those estimated by previous workers who ignored thermal runaways. The ecospheres of K0 stars are surprisingly narrow, the reason being the relatively slow rate of increase in luminosity of these stars as compared to the Sun. A planet far enough from a K star to avoid a runaway greenhouse effect early in its history will fall victim to a runaway glaciation at the time of the first appearance of an oxidizing atmosphere. There is absolutely no ecosphere at all about most K stars nor about any M stars. The ecospheres of stars more massive than the Sun are correspondingly wider. However, stars more massive than F5 increase in brightness so rapidly that any planet close enough to be able to escape runaway glaciation when an oxidizing atmosphere appears will unfortunately succumb to a runaway greenhouse effect very early in its history. The unexpectedly narrow ecospheres of sunlike stars seems to indicate that life-supporting planets could be quite rare in the galaxy; there is the distinct possibility that we could actually be alone.

Planets Within Ecospheres

We can get some quantitative idea of just how rare life-supporting planets should be by estimating the probability that a given sunlike star will have a planet situated within its ecosphere. The narrower the ecosphere, the smaller the probability that a planet will happen to form inside it. The probability that a planet will condense in a given region of space is, of course, determined by the mechanism by which planetary systems are actually formed, a process of which we know very little. However, we can get some crude idea by looking at our own solar system. Here, the planets have all formed in roughly circular orbits spaced at more or less regular intervals. The orbits of the Sun's inner four planets are about 0.4 AU apart. Although there is still no adequate theory of planetary formation, orbital spacings in extrasolar planetary systems should be similar. The probability that a planetary orbit will fall within the ecosphere of a given sunlike star can be roughly estimated by dividing the width of the ecosphere by the average orbital spacing. This has been done in the last column in Table 2-3. The probability that there is a planet within the ecosphere of a G2-type star (identical to the Sun) is only 0.13. Only thirteen out of a hundred G2 stars have planets within their ecospheres. This estimate contrasts sharply with some earlier, more optimistic projections (e.g., by Frank Drake and Carl Sagan), which suggested that each sunlike star in the galaxy could have as many as two planets capable of supporting life. The formation of the Earth in its current location vis-à-vis the Sun seems to have been a rather improbable event, possibly repeated only rarely in the galaxy. The probabilities that stars less massive than the Sun have planets inside their ecospheres are even smaller, dropping to zero for stars of classes K and M. If the probabilities listed in Table 2-3 are averaged according to the relative populations of stars of a given class, the fraction of all sunlike stars (designated by f_{ec}) that have planets orbiting inside their ecospheres can be determined. This turns out to be:

$$f_{ec} = 0.07$$

Only seven out of a hundred sunlike stars (class F5 to K9) have planets within their ecospheres. This is a surprisingly small number, indicating that worlds like Earth might be quite hard to find in the galaxy.

THE PROPER PLANETARY MASS

Even if a given sunlike star happens to have a planet orbiting within its ecosphere, that particular planet might happen to be either too small or too large to be able to support life. A planet that was too small could not retain an atmosphere sufficiently dense to protect its surface against temperature extremes fatal to life. Liquid water would all freeze solid to the surface or else boil off into the atmosphere. Because of its thin atmosphere, the planet would lack effective shielding against its star's dangerous ultraviolet light. Mercury and Mars seem to fall within this "too-small" category. On the other hand, if the planet happened to be too large it would gather up an excessive amount of hydrogen, helium, methane, ammonia, and water vapor from the protosolar nebula, forming a massive crushing atmosphere of reduced gases. In our solar system, this seems to have happened in the cases of Jupiter, Saturn, Uranus, and Neptune. All of these worlds have rocky cores that are significantly more massive than the Earth, as large as 28 Earth masses in the case of Jupiter. The dense cores of Jovian-type planets probably resulted from an initial accumulation of rocky and/or metallic planetesimals in the outer solar system 4.7 million years ago. This accumulated mass happened to be large enough to produce a catastrophic instability in the surrounding protosolar nebula, causing a massive amount of hydrogen and helium to collapse down to form the gas giant planets currently seen. These worlds retain the original reducing atmospheres that they acquired from the solar nebula. The temperatures and pressures in the atmospheres of such Jovian-type planets are undoubtedly far too high for life forms to survive. In addition, sunlight will be prevented from reaching deep into the atmospheres of such massive worlds by the dense clouds, making any sort of photosynthesis impossible.

In our own solar system, all of the planets except Venus and the Earth itself appear to be either far too small or much too large. Earth was fortunate in being large enough for the maintenance of a dense, life-supporting atmosphere, but not so large that it gathered an excessively massive reducing atmosphere from the solar nebula. How much more massive could the Earth have been before its atmosphere would have been too dense to support life? How much smaller could the Earth have been before its atmosphere would have been too thin? No precise estimate can be made, but Stephen Dole has suggested that the limits on the mass of any life-supporting planet are between 0.40 and 2.35 Earth masses. Planets with masses lying within these bounds are said to be *earthlike*. In our solar system only Venus, and of course the Earth itself, satisfies this criterion. If planetary masses in other solar systems are similarly distributed, the probability that a planet within the ecosphere of a given sunlike star is of the proper mass to support life is only 2/9 = 0.2. This is probably an overly pessimistic estimate, as there seems to be a sharp distinction between the smaller inner terrestrial planets and the outer gas giants with far more massive rocky cores. If giant Jovian-sized planets never appear in the inner parts of the solar system, then only the inner terrestrial planets of our solar system should be considered in the estimate. If so, then two out of four planets in the inner part of a given planetary system can be expected to be earthlike. The probability that a particular planet orbiting within the ecosphere of a sunlike star is earthlike then rises to 2/4 = 0.5. We take the fraction of planets orbiting within the ecospheres of sunlike stars that are earthlike (designated by f_{earth}) as:

$$f_{earth} = 0.5$$

THE RIGHT EQUATORIAL INCLINATION

Suppose that the Earth had appeared in its present location in the solar system in much the same state as it is now, only with an equatorial inclination much steeper than its current 23.5 degrees. Perhaps it could have had an inclination as large as 90 degrees. The northern hemisphere would then be in perpetual daylight for six months out of the year and in perpetual darkness for the other six. Temperatures might be far too hot for one half of the year and far too cold for the other half. Such conditions might be too severe for life. Stephen Dole estimates that the equatorial inclination of an earthlike planet should never exceed 80 degrees in order for at least 10 percent of its surface area to be habitable. Only the equatorial inclination of the planet Uranus falls outside these bounds; all the other planets have rotational axes that are more or less perpendicular to their orbital planes. Eight out of nine of the planets in our solar system therefore have orbital inclinations consistent with the development of life, although all but one fail other crucial tests. If the evolution of extrasolar planetary systems proceeds along lines similar to our own, the fraction (designated by f_{eq}) of the earthlike planets within the ecospheres of sunlike stars that have equatorial inclinations that are not too steep is equal to:

$$f_{eq} = 8/9 = 0.9$$

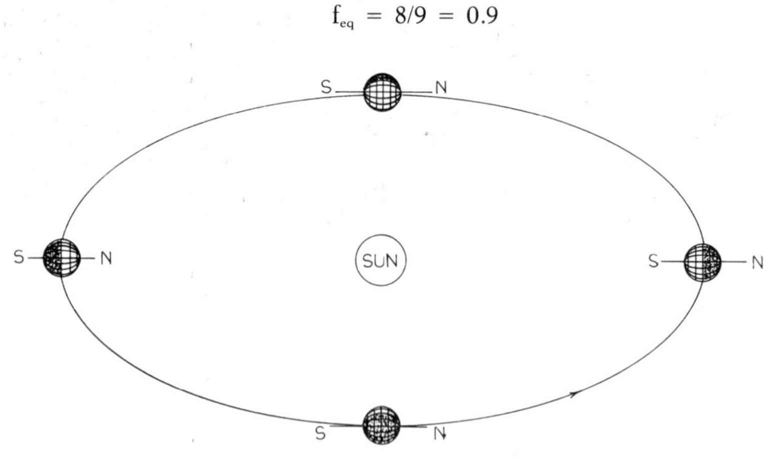

FIGURE 2.11. The orbit of Uranus, a planet whose equator is perpendicular to its orbit.

THE CORRECT ROTATION RATE

It just might happen that a planet in another solar system meets all the habitability criteria described above: its star is of the right spectral type, it orbits within the ecosphere, it has the correct mass, and it even has an equatorial inclination that is not too steep. This planet may nevertheless be rendered lifeless and barren by having a spin rate which is either too fast or too slow. If the Earth turned much more slowly, the nighttime and daytime periods would be much longer and unacceptable extremes of temperature might appear, much like those on the Moon. If the Earth were spinning much faster than it is at present, its surface might perhaps have a much more violent weather pattern, one so extreme that it would have killed any incipient life forms before they had a chance to take hold and propagate. Stephen Dole estimates that the upper limit for the length of the day on an earthlike planet is 96 hours (4 Earth days) and that the lower limit is 2 to 3 hours, although there is considerable room for error. Only Mercury, Venus, and Pluto have rotation rates slower than 96 hours. Mercury has apparently been slowed by tidal coupling with the Sun, an effect which has already been considered in the calculation of f_{sun}. Venus has a rotational period of 243 days and does not appear to be tidally coupled to the Sun. The abnormal length of the Venusian day may have played a role in the extraordinarily high surface temperatures on that planet. Pluto seems to have a rotational period of about 6.4 days, but the presence of its nearby moon Charon may have played a role in braking its rotation. No planet in our solar system has a day any shorter than about 10 hours. If extrasolar planetary systems evolve along similar lines, about 2 out of 9 earthlike planets otherwise capable of supporting life will be rendered uninhabitable by spinning too slow. So the fraction (f_{rot}) of all earthlike planets that have rotation rates that are neither too fast nor too slow can be taken as:

$$f_{rot} = 7/9 = 0.78$$

HABITABLE PLANETS

A *habitable planet* is defined as one upon which living beings not unlike terrestrial organisms could eventually evolve. The fraction of all the stars in the galaxy that have habitable planets (designated by f_{hab}) is simply the product of the individual probabilities for the presence of conditions favorable for the evolution of life, namely

$$f_{hab} = f_{sun}\, f_I\, f_p\, f_{ec}\, f_{earth}\, f_{eq}\, f_{rot} = 0.0020 \qquad (2\text{-}1)$$

The most severe limiting factor in this equation is f_{ec}, the term which takes into account the unexpectedly small sizes of the ecospheres of sunlike stars. The analysis described here is quite naturally based on the assumption that other planetary systems have evolved along lines quite similar to our own. This is sometimes called the *principle of mediocrity,* as it proposes that there is nothing particularly special about the Earth and its place in the universe. The same laws of physics should apply everywhere in the galaxy and should produce planetary systems that look very much like our own, with only random, statistical variations in such details as planetary masses, orbital spacings, spin rates, and orbital inclinations.

At first sight, the prospects of less than one in five hundred star systems being capable of supporting life is somewhat disappointing. However, there are 200 billion stars in the galaxy. As many as 400 million of them might ultimately be capable of producing living beings not unlike ourselves. However, even though the number of stellar systems with life may be large, the galaxy is a huge place, and the mean distance between habitable planets should be about 25 light-years. The search for life forms like ourselves will be long and arduous.

CHAPTER THREE

GALACTIC LIFE AND CIVILIZATION

It is possible that nearly 400 million planets in the galaxy are suitable for habitation. What are the chances that any of them have actually evolved living creatures? Do those planets which have life inevitably produce intelligent creatures somewhat like ourselves, or are we alone in the galaxy? Do other intelligent creatures look and think like us, or are they totally alien in appearance? Is any meaningful communication with them possible? What are the chances that beings somewhere else in the galaxy have produced technological civilizations similar or superior to our own? What would these societies be like? Such considerations are the subject of the present chapter.

THE APPEARANCE OF LIFE ON HABITABLE PLANETS

What are the chances that life actually does appear on a planet where the conditions are favorable? Perhaps the appearance of life on Earth was the result of a long series of extraordinarily improbable coincidences, events so rare that they are repeated nowhere else in the galaxy. The galaxy may be filled with earthlike planets that are entirely sterile and barren.

The Stuff of Life

The basic biochemistry that governs living things on Earth will perhaps shed some light on the possibility for life on other habitable planets. All terrestrial organisms, no matter how complex, are constructed of simple building-block molecules. The building-block molecules are primarily amino acids, mononucleotides, and sugars. All living things, from simple protozoan to human, use exactly the same set of building-block molecules.

The *amino acids* are the basic structural units which form proteins and enzymes within the cell. There are only twenty different amino acids commonly found in living creatures. *Proteins* are long chains of amino acids. The order in which the individual amino acids appear on the chain uniquely determines the properties of the protein, particularly the three-dimensional shape. Those proteins which can act as catalysts for biochemical reactions are termed *enzymes*.

The *sugars* or *carbohydrates* have the basic chemical formula $(CH_2O)_n$, where n is an integer. The sugars are important as basic sources of fuel, providing the energy required for the biological

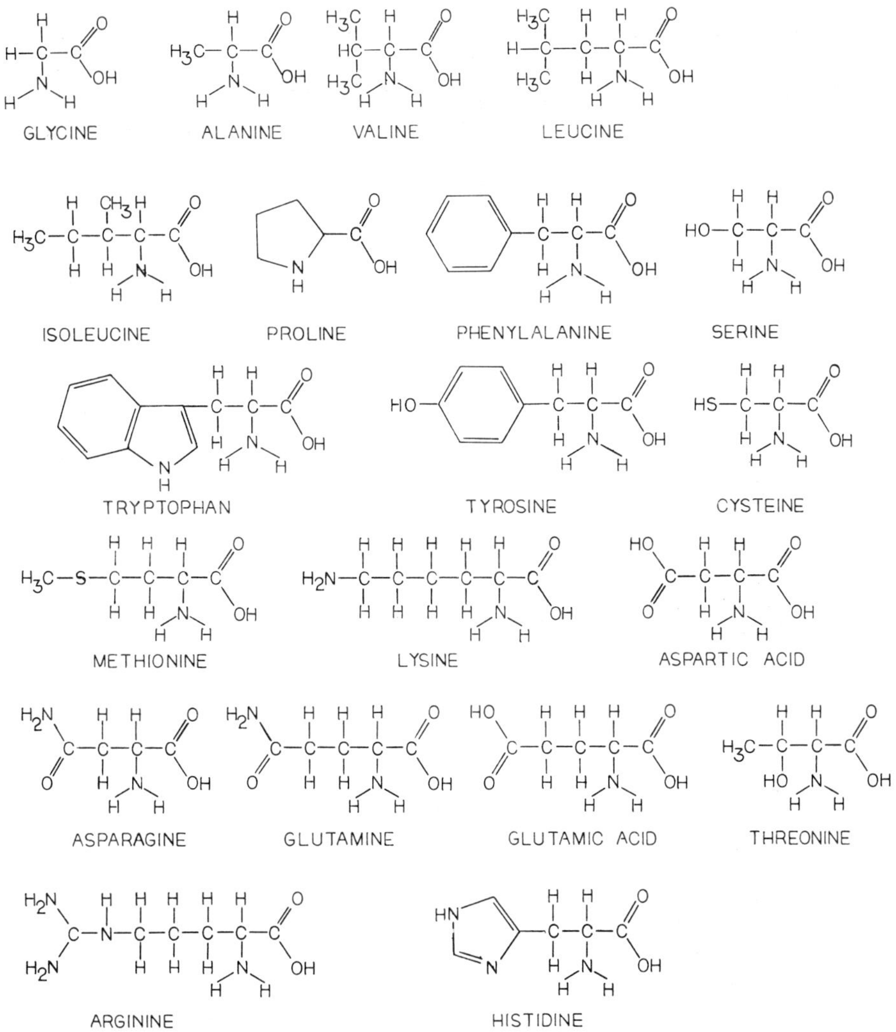

GLYCINE ALANINE VALINE LEUCINE

ISOLEUCINE PROLINE PHENYLALANINE SERINE

TRYPTOPHAN TYROSINE CYSTEINE

METHIONINE LYSINE ASPARTIC ACID

ASPARAGINE GLUTAMINE GLUTAMIC ACID THREONINE

ARGININE HISTIDINE

FIGURE 3.1. The twenty amino acids.

processes within the cell. They are also important structural components. The most abundant carbohydrate in living beings is glucose. Another important carbohydrate is ribose, a significant component of nucleic acids.

With the exception of glycine, all the amino acids can exist in two different three-dimensional forms (labeled D and L) that are mirror images of each other. These forms are known as *stereoisomers;* they have the same chemical formula, but a different spatial arrangement of atoms. All the amino acids found in living creatures belong to

GLUCOSE

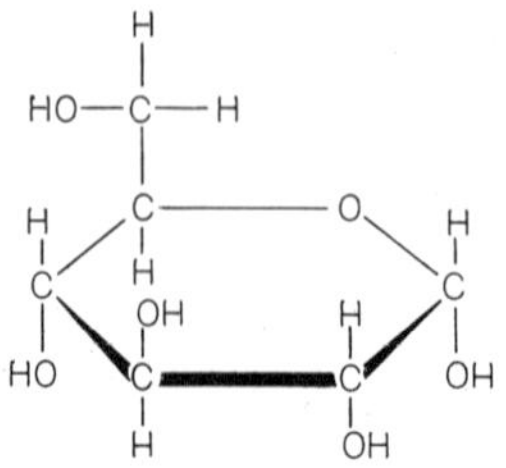

RIBOSE

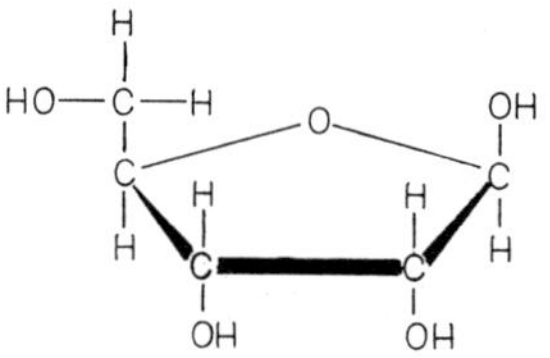

FIGURE 3.2. Glucose and ribose.

D-ALANINE

L-ALANINE

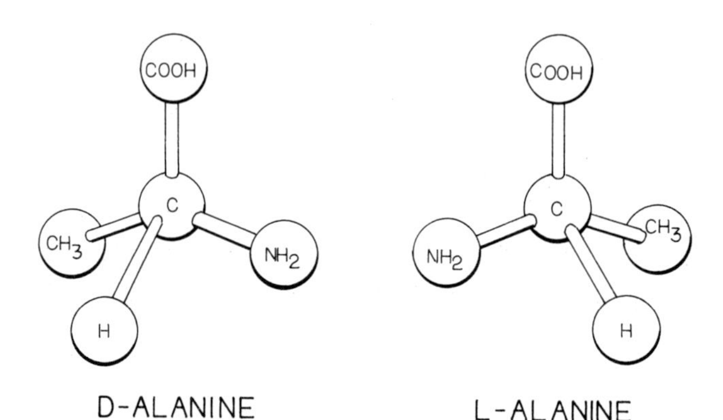

FIGURE 3.3. D and L stereoisomers of alanine.

the L stereochemical series. Most carbohydrates also have different stereoisomers, but only D sugars are found in life.

A *nucleotide* is a molecule consisting of a nitrogenous base, a ribose sugar, and a phosphoric acid. There are only five bases commonly found in living creatures: *uracil, thymine, cytosine, adenine,* and *guanine.* An important nucleotide is adenosine triphosphate, better known as *ATP.* The ATP molecule is the primary means by which energy is transferred from one place to another within the cell.

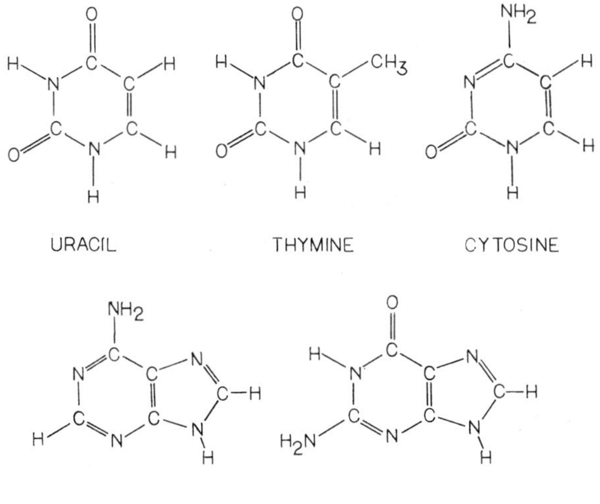

URACIL THYMINE CYTOSINE

ADENINE GUANINE

FIGURE 3.4. Nitrogenous bases.

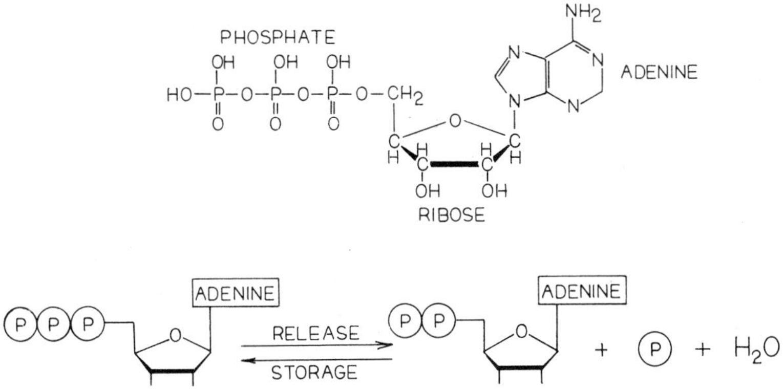

PHOSPHATE ADENINE RIBOSE

ADENINE RELEASE STORAGE ADENINE + P + H_2O

FIGURE 3.5. Adenosine triphosphate and the ATP-ADP energy cycle.

The Message of Life

Nucleic acids are long chains of nucleotides. They direct the biochemical processes within the cell and are involved in the transfer of genetic information when the cell divides. Deoxyribonucleic acid (better known as *DNA*) is the carrier of all the genetic knowledge within the cell. It consists of two separate nucleic acid chains coiled around each other to form a double helix. The order in which the base pairs appear on the DNA strand determines the order in which amino acids are placed on a protein molecule. The relationship between the base order on the DNA molecule and the amino acid which is specified is called the *genetic code.* All organisms, from bacteria to human, have exactly the same genetic code, suggesting a common origin for all life. The particular segment of a DNA strand which codes for a single protein molecule is called a *gene.*

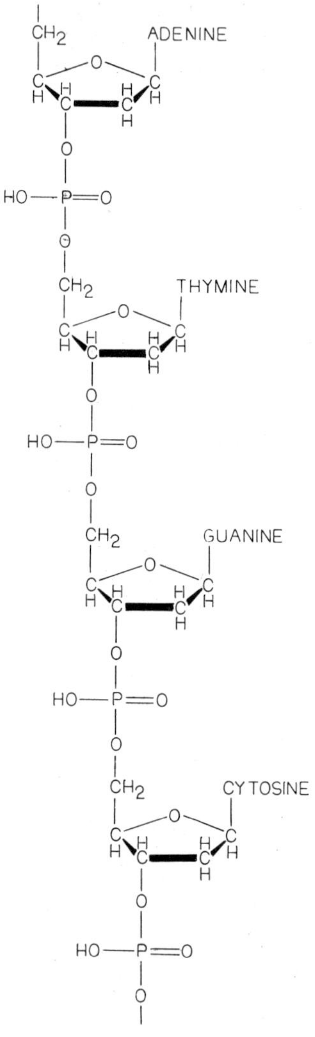

FIGURE 3.6. DNA.

The two strands in a DNA double helix are complementary to each other; they each contain the same base-pair sequence information. When it comes time for a cell to reproduce by dividing into two, the two strands of the DNA unwind. Each strand then acts as a "template" for the synthesis of a complementary strand of DNA. When this process is completed, there are two identical daughter strands of DNA, each containing one strand of the parental DNA. One of the new DNA molecules goes to each of the cells, transmitting the knowledge essential for protein manufacture to the descendants.

Alterations in the structure and/or base-pair sequence of the DNA that take place in such a way that a different protein is produced are called *mutations*. A cell that has undergone mutation will synthesize different proteins than its parents did and will possess different characteristics. In most cases, these changes are disadvantageous, and the cell dies before it has a chance to reproduce. In rare cases, the changes are favorable, and the mutant has a better chance to survive than its parents. It will pass its newly acquired characteristics on to its descendants when it divides.

Spontaneous Generation?

The probability that other habitable planets will evolve life forms is related to the process by which life came to exist on Earth. Until quite recent times the answer to this question was regarded as unknowable by mortal men, the creation of life on Earth having been performed by the miraculous intervention of supernatural beings. According to the Bible, human beings as well as all living creatures were created by God. However, most people also accepted the idea that some lower forms of life could appear spontaneously within inanimate matter without the intervention of divine beings. This was imagined to happen when maggots suddenly appeared inside rotting meat.

In 1668, the Italian physician Francesco Redi was probably the first scientist to cast serious doubt on the possibility of the spontaneous generation of living beings. He noticed that if flies were kept away from the meat maggots never appeared, no matter how long the meat was exposed. Redi correctly concluded that the maggots had come from microscopic eggs laid on the meat by the flies. When microscopic organisms were discovered, it was at first assumed that they were produced by spontaneous generation. A fresh broth would soon be contaminated with microscopic organisms, even if flies were excluded. However, in 1765 the Italian naturalist Lazzaro Spallanzani demonstrated that a broth which had been presterilized would never become contaminated, so long as it was not exposed to fresh air. Even in the face of this evidence, however, some people still felt that the sterilization process itself destroys the "vital force" within the air over the broth, preventing the formation of living creatures. It took Louis Pasteur to finally lay the notion of spontaneous generation to rest in 1862 by carefully filtering unsterilized air before admitting it to a previously sterilized broth. The broth remained sterile, demonstrating that the air itself could not be the cause of the growth, but rather bacteria carried by the air into the broth from the outside.

Life from Outside?

If spontaneous generation of living creatures is to be ruled out, are we then forced by default to accept a supernatural explanation for the origin of life? In 1907, the Swedish chemist Svante Arrenius published a book entitled *Worlds in the Making,* in which he proposed that life had always existed in the universe and had migrated across interstellar space to the Earth. It traveled in the form of small, dormant spores that escaped from some distant planet and were driven out into interstellar space by the pressure of light from that planet's central sun. The spores drifted through space for countless eons until a few of them landed on the barren Earth. The spores then sprung to life, and the process of evolution began. Theories of this nature are often referred to as "panspermia," as they teach that life will inevitably spread throughout the entire galaxy to all those planets capable of sustaining it.

This theory initially sounds quite attractive, but it can be subjected to severe and perhaps fatal criticism. First, it is now known that stars emit intense ultraviolet radiation, as well as occasional bursts of X-ray and gamma radiation. Even the most hardy spore is likely to receive such a large dose of radiation during its travels through space that it would be killed. Second, the same light pressure responsible for driving the spores away from their sun would prevent them from entering our solar system. The pressure of sunlight would push them right back into interstellar space.

The attractive idea of "panspermia" dies hard. In 1973, Francis Crick and Leslie Orgel introduced an even more daring variant of this old theme. They proposed that the first life was brought to Earth by intelligent beings from another star system. Two arguments are used to support this hypothesis. One is derived from the abundances of trace elements within living creatures. Molybdenum is essential for life, whereas nickel appears to be relatively unimportant. Molybdenum is rather scarce in Earth's crust, and its importance in living organisms could mean that they originally evolved in an environment much richer in this element. Other explanations are certainly possible; molybdenum may turn out to be chemically irreplaceable in certain essential biochemical reactions. Molybdenum is, in fact, reasonably abundant in sea water, where life first became established. Crick and Orgel also use the universality of the genetic code as evidence for an infective theory of the origin of life. It is initially rather surprising to find that organisms using slightly different genetic codes do not exist. It may have happened that organisms using many different codes lived side by side in the primitive oceans but that the present code was so vastly superior to all others that it displaced them. The notion of "directed panspermia" is certainly provocative, but much more work is needed before it can be regarded as anything more than an unsupported hypothesis.

Fred Hoyle and Chandra Wickramasinghe of the University College at Cardiff in Great Britain have recently suggested that the first life may have appeared inside a comet rather than on the Earth itself. A comet is made of ices of water, methane, and ammonia intermixed with silicate dust and metallic debris. There are also more complex organic molecules present, some of which may be of biological significance. Comets which come close to the Sun undergo periodic heating cycles in which they are progressively stripped of their volatile elements, perhaps leaving behind a concentrated "soup" rich in organic material dissolved in liquid water. As this mixture was warmed by the early Sun and irradiated with ultraviolet light, the ingredients of the soup could have formed progressively more complex organic molecules. Given enough time this process might perhaps have led to the formation of primitive life within the comet. These organisms were brought to Earth when the comet collided with the young Earth, and a few of the hardier varieties took hold and propagated within the primeval seas.

Is there any evidence for the initial formation of life within a comet? A rare type of meteorite (the carbonaceous chondrite) is rich in organic compounds, some of which appear to be similar to those found within living beings on Earth. Carbonaceous chondrites may be the residues of old comets that were long ago stripped of their volatiles. Most are approximately 4.5 to 4.7 billion years in age and must date back nearly to the origin of the solar system itself. The first known fall of a carbonaceous chondrite meteorite took place near the town of Alais in France in 1806. The well-known Swedish chemist J. Jakob Berzelius examined this meteorite in 1834 and found that its interior looked like humus. Since that time about twenty additional carbonaceous chondrites have been recovered, and organic material has been found in many of them. The Orgueil meteorite of 1864 contained paraffins, long-chain aromatics, hydrocarbons, fatty acids, and porphyrins. Some later meteorites contained amino acids, some of which were identical to those present in terrestrial organisms. Bartholomew Nagy and George Claus found the Orgueil meteorite to be rich in "highly structured forms" similar in appearance to fossil terrestrial microorganisms. Some workers, including the Soviet scientists Bairiev and Mamedov and the American Frederick Sisler, have claimed to have found living organisms in meteorites!

The claims for the presence of living organisms inside carbonaceous chondrite meteorites must be assayed with caution, as it is extremely difficult to prevent contamination by terrestrial organisms. It has been shown that even carefully sterilized meteorites become contaminated with microorganisms simply by sitting on the shelf. Edward Anders and Frank Fitch, of the University of Chicago, noted that the "highly structured forms" noted by Nagy and Claus in the Orgueil meteorite were remarkably similar to ragweed pollen. Since outer space is unlikely to be populated by ragweed, terrestrial contamination is the most likely explanation. Cyril Ponnamperuma and his group at the University of Maryland have found that many of the organic molecules present in carbonaceous chondrite meteorites are not of a type found in any terrestrial organisms. In addition, amino acids of both D and L varieties are found. The organic molecules within these meteorites undoubtedly had an extraterrestrial origin, but it is also almost certain that they were formed entirely abiologically.

There is no credible evidence for an extraterrestrial biochemistry ever having taken place within comets or meteorites. Short-period comets that pass close to the Sun generally last only a few thousand years before they are completely stripped of all their volatiles and disintegrate into many thousands of small rocks and fine grains of debris. This time period is undoubtedly far too short to allow any sort of biological evolution to take place.

The Primordial Soup

Today, most workers believe that life originated here on Earth and not in outer space. Life was already present on Earth only a billion years after the planet condensed from the primeval solar nebula. At the time of the origin of life, the atmosphere was a mixture of reduced gases, such as carbon dioxide, ammonia, water vapor, methane, and hydrogen sulfide. The Soviet chemist Alexander Oparin suggested in 1925 that this reducing environment was absolutely essential for the origin of life. Life could not possibly have arisen within the oxidizing atmosphere of today. The British biochemist J.B.S. Haldane independently made virtually the same suggestion a few years later, but neither theory initially attracted very much attention. It was not until 1952 that any experimental evidence for the Haldane-Oparin theory was obtained. In that year Stanley Miller and Harold Urey, of the University of Chicago, constructed an apparatus that simulated an

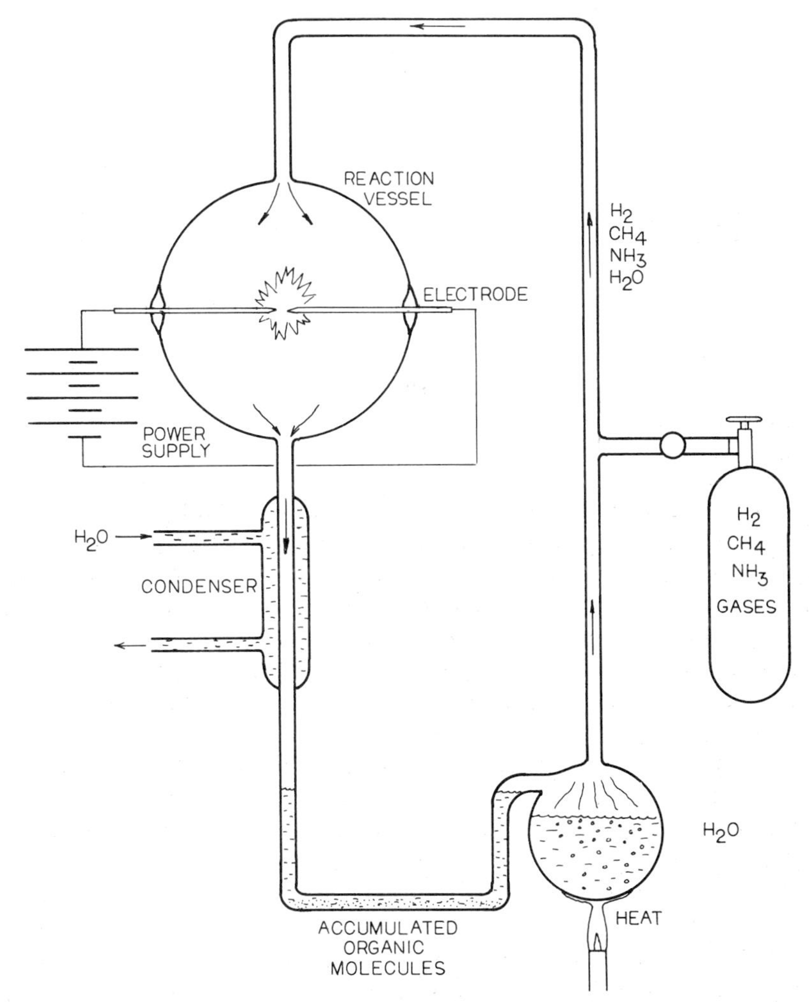

FIGURE 3.7. Apparatus used by Stanley Miller to create amino acids by passing a spark discharge through a mixture of gases simulating the primitive atmosphere.

ancient thunderstorm. A mixture of gases presumed to have been present in the primeval atmosphere was contained within a vessel and subjected to an electrical discharge. After a few days of operation a considerable amount of organic material had accumulated, including some amino acids. Subsequent experiments demonstrated that ultraviolet light, X-rays, gamma rays, high-energy electrons, alpha rays, energetic neutrons, or even supersonic shock waves would work just as well as an electrical discharge in creating amino acids from the reducing gas mixture. Although Miller and Urey used a highly reducing mixture of hydrogen, ammonia, methane, and water vapor in their early studies, later experiments by Philip Abelson of the Carnegie Institute demonstrated that a considerably less-reducing mixture of carbon dioxide, carbon monoxide, hydrogen, nitrogen, and water vapor subjected to radiation would produce amino acids just as readily. The addition of hydrogen sulfide to the gas mixture gives a rich variety of organic sulfur compounds. Other experiments have synthesized nitrogenous bases and sugars. Organic molecules are even produced when reducing gas mixtures are dissolved in aqueous solution and subjected to ultraviolet light, to ionizing radiation, or merely to heat. Widely varying mixtures of several different gas molecules have all been found to yield complex organic molecules when subjected to energetic excitation; the particular details of the gas mixture, the means of excitation, or even the physical environment are apparently not all that critical. Carl Sagan and Iosif Shklovskii have estimated that the rate of production of organic molecules in the primeval terrestrial atmosphere and oceans could have been as high as a ton per second all over the Earth. The organic material must have accumulated quite rapidly in the oceans, and there could have been as much as 1 or 2 percent organic material in the seas very soon after the first oceans appeared.

All that seems to be required for the generation of copious amounts of organic molecules is a mixture of simple gases and a source of energy. There is, however, one important exception. Numerous investigators have found that the addition of a slight amount of molecular oxygen to the experimental vessel invariably inhibits the synthesis of organic molecules. Rather than a mixture of amino acids and other organic molecules, a biologically uninteresting "smog" is created. If there had been any significant amount of molecular oxygen in the Archean atmosphere, no organic molecules would have been synthesized, and life would never have appeared on Earth.

The Chain of Life

The next step in the formation of life must have been the joining together (or *polymerization*) of the molecules in the primordial "soup" to form the proteins, nucleic acids, and polysaccharides that are the stuff of life. Amino acids, nucleotides, and simple sugars do not spontaneously join to form larger molecules when placed in aqueous solution. Some sort of external "linking" agent is required. In present-day organisms this feat is accomplished by the combined action of ATP and complex systems of enzymes, but these were of course not present in the primitive oceans. Another problem is presented by the fact that a Miller-Urey type of experiment also produces lots of organic molecules that are not related to any used by life. Why did nature discard them? Finally, D and L stereoisomers are created in equal amounts when organic molecules are generated artificially. Why did nature select only L amino acids and D carbohydrates for life?

It has been suggested that inorganic phosphate ions dissolved in the water could have played the role of ATP in initiating biologically important polymerizations. Scientists at the Max Planck Institute for Virus Research, in West Germany, were able to use polyphosphate esters to form protein chains of up to 24 amino acids in length. Some primitive organisms in fact do store their excess energy in the form of inorganic phosphates rather than ATP, which may be reminiscent of phosphate-dependent polymerization processes that took place in the distant past.

Another possibility is that ATP could have been one of the earliest polymeric molecules to have formed. Ponnamperuma and Kirk were able to produce ATP by subjecting an aqueous solution of adenine, ribose, and ethyl metaphosphate to ultraviolet light. If ATP could have been formed by this method, it could have acted as a coupling agent in the formation of even more complex nucleic acids and proteins.

An entirely different approach involves the consideration of processes by which amino acids, nitrogenous bases, phosphates, and sugars in the primitive oceans could have become more highly concentrated or else removed from water altogether, so that polymerization reactions could take place. The chemist J. D. Bernal, of the University of London, suggested that organic molecules could have been absorbed onto clays and held together long enough to polymerize. Akabori and his co-workers have polymerized aminoacetonitrile in the presence of acid clays to form glycine proteins. The chemist Aharon Katchalsky, of the Weizmann Institute of Science in Israel, has demonstrated that clays can catalyze the polymerization of amino acid adenylates to form a proteinlike chain of up to 50 units long. It has recently been found by Stephen Bondy and Marilyn Harrington, of the University of Colorado, that certain clays preferentially bind L-type amino acids rather than D-type, suggesting a process by which proteins with L-type amino acids became dominant.

The First Cells

Sidney Fox, of Florida State University, found that dry mixtures of amino acids will spontaneously polymerize when heated to 130 degrees Celsius. The addition of inorganic phosphates lowers the polymerization temperature to 60 degrees Celsius. Such techniques can produce protein chains of up to 200 units long. These artificially created chains are called *proteinoids*. Billions of years ago the primitive oceans must have washed onto the slopes of active volcanic cones. Some of the water collected in low-lying areas and was evaporated by the heat, leaving behind a dry film of organic molecules. These were subsequently polymerized by the heat, and the resulting proteinoids were washed back out into the oceans. Fox found that proteinoids heated in concentrated aqueous solution will spontaneously aggregate into small, micron-sized spheres. Some of these microspheres develop a sort of outer boundary layer that shields the interior from water. A few even have a weak catalytic activity that can mimic biological reactions, such as the decomposition of glucose.

Alexander Oparin has found that an aqueous solution of proteins and polysaccharides will spontaneously form droplets known as *coacervates*. The polysaccharides act as a membrane enclosing a protein-rich interior. If left to themselves, coacervates quickly disperse. They can be made to last longer by adding the enzyme phosphorylase, which gathers within the interior of the coacervate. If glucose-1-phosphate is added to the solution, some of the sugar molecules are enzymatically attached to the membrane by the action

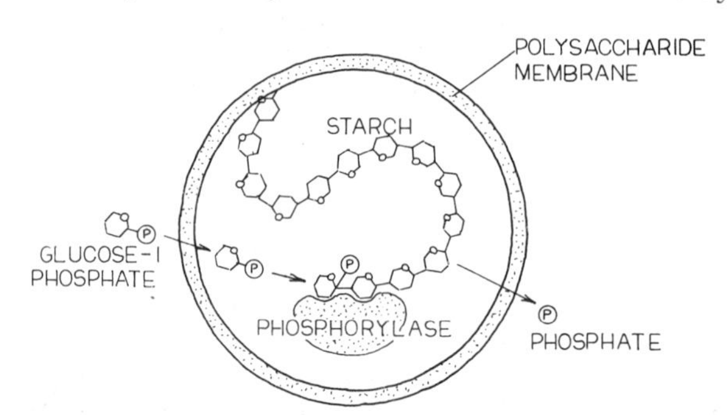

FIGURE 3.8. A coacervate droplet with phosphorylase inside will grow by enzymatically attaching glucose molecules to the starchy membrane. From Dickerson (1978). Copyright © 1978, *Scientific American*, Inc. All rights reserved.

of the phosphorylase. A coacervate droplet with phosphorylase inside will therefore grow by absorbing nutrients from the outside, mimicking one of the functions of a living cell. Droplets even split into two when they grow too large, but daughter droplets do not grow any further unless more phosphorylase is added. If some way could be found for the droplets themselves to make the phosphorylase, the coacervates would be self-perpetuating "organisms" with a one-step metabolism.

A picture of the origin of the first living cell now emerges. The primitive oceans must have accumulated a large number of coacervate droplets or thermal proteinoids, each possessing special characteristics. Those droplets that by chance had the ability to catalyze "useful" reactions stabilized and grew, whereas those that did not quickly dispersed. There was a strong selective advantage for those droplets that had the ability to take energy and building-block molecules from their surroundings and incorporate these into structures that promoted growth and prevented dispersal.

Such a growing, self-perpetuating droplet is not exactly living matter, but is very close to it. One essential element is lacking. It needs some means by which its peculiar characteristics can be passed on to its descendants when it divides. How was the first gene formed? Which came first, the protein or the gene? Leslie Orgel proposes that the protein came first. By chance, a few prebiotic protein-rich coacervate droplets appeared which were able to bind nucleotides with some selectivity. Once bound to a protein chain the nucleotides could have been polymerized into a nucleic acid chain, the base order on the chain being complementary to the amino acid order on the protein. The first DNA molecule may have formed in this manner. J.B.S. Haldane has suggested that the gene may have come first. The first living organism may have been a "naked gene," being little more than a large DNA molecule that could make copies of itself without the assistance of enzymes. The acquisition of a system of enzymes and a cell boundary could have come later.

Life in the Cosmos

Is it possible that the sequence of events which produced the first life on Earth has been repeated in numerous variations elsewhere in the universe? Radio astronomy has found that there are large amounts of ammonia, methane, carbon dioxide, and water vapor concentrated in the large clouds of gas and dust which form new stars. Their presence has been detected by measuring the radio noise that they emit. Earthlike planets should have primitive atmospheres rich in these gases. The stuff of life should be readily synthesized in vast amounts in the atmospheres and oceans of these worlds by sources of energy available almost everywhere. The oceans of newly formed, earthlike planets should be filled with organic molecules. Given enough time, assemblies of molecules having the properties of life should inevitably form wherever the conditions are favorable. If the "assumption of mediocrity" is valid, life should eventually arise on all planets that can possibly support it. The fraction (designated by f_{life}) of all habitable planets that ultimately evolve indigenous life can be set equal to unity:

$$f_{life} = 1$$

Even though life should eventually appear on all planets that are capable of sustaining it, there is no reason to expect that the biochemistry of beings living on other worlds should be exactly the same as that of terrestrial organisms. Even though all life should be based on carbon, nitrogen, hydrogen, and oxygen, there are nevertheless myriads of different biochemical pathways which nature could choose in evolving living beings. There is no reason why extraterrestrial creatures should have to use exactly the same twenty amino acids as Earth organisms, or even that they must use amino acids at all. The means by which genetic information is passed on to descendants may be entirely different from that of terrestrial life. The genetic code may be different, involving bases other than adenine, thymine, guanine, or cytosine. The genetic apparatus may not involve nitrogenous bases at all, using perhaps amino acids or even sugars instead. Perhaps the genetic apparatus of extraterrestrial organisms does not involve any molecules similar to DNA at all. The metabolic pathways by which food is converted into matter and energy may be entirely different. The biospheres of other planets are in all probability highly toxic to our form of life, and any landings by humans upon other inhabited worlds might result in the instant death of everyone aboard the spaceship. Nevertheless, the chemistry of all living beings in the universe should be the chemistry of carbon, and the laws of biochemistry should produce the same sorts of reactions in all living things, with only the specific details differing from planet to planet.

LIFE AND INTELLIGENCE

Although the probability that any single star system has a planet with living creatures is only one in five hundred, there are so many stars in the galaxy that there could be as many as 400 million inhabited worlds. What is the probability that *intelligent* life will appear on these worlds? Intelligence is here defined as the ability of an organism to gather sensory data from its surroundings and make decisions about future action based on an analysis of this information. Intelligence is a relative attribute; men are more intelligent than dogs, dogs are more intelligent than fish, and fish are more intelligent than clams. Nevertheless, all of these creatures have one thing in common: They have complex sensory organs which are connected by a nervous system to a central brain. Unlike simpler organisms, they are able to adapt to changing circumstances. On Earth, the advent of sensory and nervous organs was made possible by the evolution of multicellular life.

Multicellular Life

It seems quite reasonable to presume that the appearance of metazoan life is a prerequisite for the evolution of intelligence. What then are the chances that other worlds can develop multicellular life forms? Primitive, one-celled organisms may appear on most earthlike planets where conditions are favorable, but the evolution of more complex multicellular life may be so difficult that few worlds ever produce creatures larger than microbes. Single-celled life appeared on Earth almost immediately after the oceans first condensed, but it took almost 3 billion more years for the first multicellular organism to evolve. Life may actually be abundant in the galaxy, but large creatures and conscious beings such as ourselves may be relatively rare.

Multicellular life was slow in coming to Earth not because its evolution was intrinsically difficult but because an oxygen-rich environment was required. This was made possible by oxygen-generating photosynthesis. The advantages of aerobic photosynthesis are so obvious that it seems rather odd that it took a billion and a half years of painful evolution before it appeared. It was probably necessary to wait until the oxygen level had reached a relatively high level to permit the molecules necessary for aerobic photosynthesis to be synthesized. Furthermore, the evolutionary process had to be allowed time for the development of effective biochemical systems to protect against the adverse effects of oxygen before any types of oxygen-evolving creatures could possibly survive and prosper.

The evolution of multicellular life and intelligence will be impossible anywhere in the galaxy in the absence of oxygen. The probability that other inhabited, earthlike worlds in the galaxy develop oxygen-rich atmospheres depends on the odds favoring the development of aerobic photosynthesis. If terrestrial experience is any guide, oxygen-generating photosynthesis probably will take a long time to appear on each of the inhabited worlds in the galaxy. However, because photosynthetic organisms capable of using water as a prime reactant are at such a vast competitive advantage, oxygen-producing photosynthesis should ultimately appear wherever life exists. Organisms employing aerobic photosynthesis should rapidly proliferate, quickly filling the atmosphere of their world with oxygen.

Once multicellular life has appeared, the advance toward intelligence should be relatively swift. The development of higher intelligence is perhaps the most effective survival aid that an organism can have; creatures with more efficient brains are better at finding food and escaping from enemies and have a higher probability of living long enough to leave descendants. The laws of Darwin-Wallace evolution should be operating everywhere in the galaxy where there is life, and these pressures will tend to produce intelligent creatures on all worlds upon which life appears. The fraction (designated by f_{int}) of those life-bearing planets that ultimately evolve intelligent life forms can be set equal to unity:

$$f_{int} = 1$$

The Vagaries of Evolution

Even though all life-bearing planets in the galaxy should ultimately produce large, multicellular, intelligent life forms, there is no reason at all to expect that evolution will invariably follow precisely the same course everywhere in the galaxy that it has on Earth. Even here on this planet evolution has undergone many twists and turns in its production of such a large variety of creatures. Evolutionary pathways have depended on such random, unpredictable factors as continental drift patterns, changing climates, genetic mutation, the waxing and waning of ice ages, and perhaps even meteorite impacts or supernova explosions. At any point in our own evolutionary history a different mutation could have occurred and an entirely different organism would have appeared. Evolution does not exhibit any central line of progression that leads causally from a protozoan to a human being. The paths that evolution follows on other worlds will almost certainly be entirely different. The Harvard paleontologist George Gaylord Simpson argues that the mutations responsible for the evolutionary process are so numerous and random that it is very unlikely that life on other planets will bear even the remotest similarity to that on our own. In particular, the probability that natural selection has produced other humanoid creatures similar to us elsewhere in the galaxy is so small as to approach the vanishing point. Even if highly intelligent extraterrestrials do exist, they may be so different from us that there is absolutely no chance of communication.

Some authors, such as Robert Bieri, feel that there must of necessity be some gross similarities between extraterrestrial life forms and those that have appeared on Earth. It is likely that all life originates in the oceans, and there should be strong pressures there for the production of animals that are sufficiently well streamlined so that they can travel swiftly through the water in pursuit of prey or in flight from enemies. The torpedolike shape of the fish seems to be ideal for this purpose. Placement of sensory organisms and a mouth in the forward part of the animal are an obvious advantage. The constant struggle for survival in the water will lead to the development of a larger, more complex nervous system that makes the creature more clever and capable. Considerations of efficiency should lead to the placement of the central part of the nervous system near the sensory organs that provide the information needed for survival. There is also strong pressure for the presence of grasping organs near the mouth to aid in the devouring of prey. Evolutionary pressures in the oceans of other planets should produce a streamlined animal with sensory organs, a mouth, and a brain all grouped together in a head in its forward part. The movement onto land will produce different types of evolutionary pressures. Of all possible means of locomotion, walking on legs is perhaps the most efficient. Animals that are able to walk on legs have much more speed and maneuverability than those that can only slither. Wheels or any other sort of rotation as a prime means of locomotion are probably unlikely, perhaps because of the large amount of friction that would be generated by a rotating biological joint. There are also strong pressures to reduce the number of multiple appendages and to minimize the number of legs. The development of binocular vision and binaural hearing (two eyes, two ears) enables an animal to perceive depth. Considerations of efficiency should limit eyes and ears to two apiece. Evolutionary pressures on the continents of other planets are likely to be similar, producing intelligent animals with bilateral symmetry, an even number of legs (two, four, or six), and a head with sensory organs, mouth, and brain in its forward part. Extraterrestrial creatures would be grossly similar to animals that have appeared on this planet, but a moment's glance should be sufficient to tell the difference.

Even if there are beings on other worlds similar in overall physical appearance to human beings, there is still no guarantee that their thought processes will bear any similarity to our own. Human mental processes depend to a great extent on the way in which the human brain has developed from its evolutionary precursors. Paul MacLean, of the National Institutes of Health, has proposed that the human brain is constructed of three distinct tiers, each with its own mode of operation. This *triune brain,* as MacLean calls it, was produced by successively more and more complex layers of nerve cells accumulating during millions of years of evolution. The central part is the most primitive and is called the *reptilian complex.* It is a remnant of some of the earliest nervous systems that were evolved by the dinosaurs nearly 200 million years ago. It appears to be concerned with instinctual and ritualistic behavior and may play an important part in aggression and territoriality. Our unconscious tendency to follow leaders blindly may also originate here. Next is the

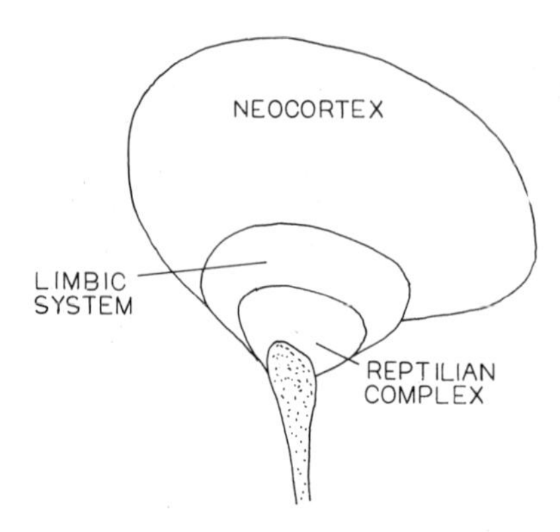

FIGURE 3.9. Paul Maclean's model of the triune brain.

limbic system, which first evolved at the time the nonprimate mammals appeared. It seems to be responsible for strong emotions, as well as for various sorts of altruistic behavior such as those exhibited in the rearing of young. The outermost layer of the brain is the *neocortex,* which was the last to appear and is most highly developed in humans. It is involved in abstract reasoning and is responsible for long-term memory. The neocortex first appeared in the primate animals and rapidly grew in size as more advanced and intelligent hominoid creatures evolved. The large neocortex separates humanity from the rest of the animal kingdom and is responsible for mankind's current dominance of the planet. The three parts of the human brain are not completely separate entities; each is in intimate contact with the others to generate the thoughts and emotions that make up the rich mosaic of human life. Different evolutionary pathways on other worlds are certain to produce radically different brain structures. Alien intelligences are likely to appear completely bizarre to us. There may be highly intelligent creatures that communicate by means of color and texture, by complex pulses of ultrasound, or by biologically produced radio waves. We may encounter beings with no concept of self, no territorial urge, no fear of strangers, no need for dominance, no religious yearnings, and no intrinsic curiosity about their surroundings. Nevertheless, these creatures will be intelligent, even though the intelligence they possess will be alien to our own.

TECHNOLOGY AND CIVILIZATION

Intelligent creatures can be expected to evolve on all planets suitable for the development of life. But how many of these intelligent life forms ever manage to produce technological civilizations similar or superior to our own? In particular, have other societies appeared in the stars that are capable of making themselves known to us on Earth, either by accident or by design? Such societies are called *communicative civilizations,* as they are capable of interstellar communication, either by direct travel or by an exchange of radio signals.

In the previous section it was argued that every habitable world in the galaxy should eventually evolve some form of intelligent life. How many of these intelligent creatures ever produce technological civilizations which we might be able to contact? Technology is an organized effort to attain control over nature in order to provide a more secure and productive mode of life. It is to be distinguished from the merely instinctual behavior exhibited by birds in the building of nests, by insects in the making of hives, or by beavers in the construction of dams. Technology implies continual change and improvement. The art of nest building by birds has advanced very little in the last 50 million years, but the art of aircraft construction by humans certainly has. The use of technology is certainly evidence that a high level of intelligence is present, but it may not necessarily follow that all intelligent beings inevitably develop technologies. After all, whales and dolphins do on occasion appear to exhibit intelligent behavior, but they have never used tools.

Life on Land

It is an interesting question as to why such intelligent creatures as whales or dolphins never produced any engineering innovations. Perhaps technology is possible only for creatures that live on dry land. Life in the water is a comparatively easy existence; the temperature is nearly constant, and there is no violent weather to disrupt the orderly and placid pattern of life. There is no need for protection against adverse climatic conditions. The high viscosity of water makes it very difficult to manipulate large objects under the seas. Another factor is, of course, that it is impossible to build a fire under water, ruling out any undersea technology based on the use of heat energy. As a result, no creature living in the sea ever developed any sort of technology.

The large-scale movement of life onto the land, which began about 400 million years ago, was made possible by the presence of an oxygen-rich atmosphere created by the evolution of aerobic photosynthesis earlier in the Precambrian. Once on the land, living creatures were subject to entirely different evolutionary pressures. Life was now under the influence of changing seasons, climatic variations, and the ever-present cycles of hot and cold, wet and dry. All of these forces acted to accelerate the processes of evolution and natural selection. Perhaps the most important influence of all was the presence of continental drift and plate tectonic phenomena. The steady movement of continental plates creates an ever-changing surface environment which enhances the rate of evolution of life forms more readily able to adapt. In populations of flora and fauna, individual specimens are suddenly separated from each other when continental plates split apart, and each subgroup then proceeds along an entirely different line of development. Entirely new species are thereby created. When continental plates collide, living beings that originally evolved completely independently of each other are suddenly forced to compete for the same territory. This competition weeds out the unfit. The moving continents produce steady climatic changes; a lush tropical jungle during one era may turn into an arid desert during the next, and new patterns of life must adapt to the change.

Continental drift may also have been essential in making the materials required for advanced technology more readily accessible to humans. The collision of massive continental plates over many eons of time created large concentrations of rich metallic ore deposits near the surface, making these materials readily accessible to humans so that they could create an economy based on the use of metals rather than stone. Had it not been for continental drift, all of the iron, nickel, copper, and zinc metals currently in the Earth's crust might have sunk to the core very early in its history or else might have been so evenly distributed and diffuse as to be virtually impossible to obtain. Without easy access to highly concentrated metallic ores, the human race would never have been able to produce tools made out of anything other than stone or wood. In addition, continental plates sliding over each other buried massive amounts of organic material which eventually formed the large deposits of coal and oil that are so essential for any technology based on the use of heat energy. Without metals and fossil fuels, the human race would never have been able to build any sort of advanced technology beyond the most primitive stone-age variety. Things such as electronics, aviation, or atomic power would have been impossible. In particular, we could never have built the spaceships or radio telescopes which may make it possible for us to communicate with other civilizations around distant stars.

Alien Technological Societies

How many other technological civilizations have appeared in the galaxy? For all we know, the stars may be populated by superintelligent creatures that are interested in philosophical contemplation or abstract mathematics rather than in acquiring domination over nature. The events which produced a technological civilization on Earth may be so improbable that they have been reproduced almost nowhere else in the galaxy. Intelligence may be abundant, but technology may be rare. On the other hand, it can be argued that the successful lifestyle that technology seems to provide makes it inevitable that every world with intelligent life will eventually evolve a technological civilization. If so, every inhabited world in the galaxy will be conquered by the first species to develop the use of tools. Technology will inevitably come to every world where the conditions are favorable.

The habitable worlds favorable to technology are those which can sustain life on dry land and which also have continental drift and plate tectonic activity. In the previous section we have argued that all worlds where life exists should eventually evolve oxidizing atmospheres. These oxygen-rich worlds should be able to support life on dry land, provided of course that the land area is not entirely submerged beneath the water. A planet perpetually covered by a global ocean obviously would never produce any land-living life, no matter how favorable it otherwise might be. However, none of the other terrestrial planets in the solar system are so flat that they would not have at least some land area jutting out of their oceans, even if they had appreciably more water than Earth. The Martian volcanoes are much higher than any mountains on Earth, and Maxwell Montes on Venus is 2 kilometers higher than Mount Everest on Earth. It is likely that the earthlike planets around other stars will also have rough surfaces with at least some dry land sticking out of the water. Large-scale continental drift on the terrestrial planets seems to be a phenomenon that is confined to the Earth alone. Valles Marineris on Mars may have been formed by the onset of a forced separation of continental plates induced by the volcanic bulge that formed the Tharsis region, but such activity did not continue for long. Perhaps Mars was too small to retain its internal heat for a sufficient length of time to sustain continental drift. Venus is about the same size and density as the Earth, and the amount of internal heating should be similar. Not enough is known about the surface of Venus to make any firm conclusions, but it now appears that its surface is very much flatter than that of Earth, with "continental" land masses occupying a very much smaller fraction of the surface area of the planet. Most planetary geologists suspect that the crust of Venus is significantly thicker than Earth's, so thick that it never broke up into mobile plates. Venus seems to be a "one-plate" planet that has been much less geologically active over most of its history than has been Earth. Much of its surface may date back to the days of intense meteoric bombardment over 4 billion years ago.

Of the two Earth-like planets in our solar system, only one has enough geological activity to maintain the continental drift essential for providing the raw materials that make high technology possible. If we adopt the notion that other solar systems evolve along similar lines, we would expect that statistically only half of all earthlike planets would have surfaces continually reshaped by continental drift activity. That is, we can take the fraction (designated by f_{tech}) of the planets with intelligent life which ultimately develop a technological society to be equal to one-half:

$$f_{tech} = 1/2$$

The fraction of all stars which ultimately develop technological civilizations (designated by f_{civ}) can now be calculated. It is simply the fraction of star systems which have habitable planets, multiplied by the fraction of such worlds which evolve living beings, multiplied by the fraction of inhabited worlds which evolve intelligence, and finally multiplied by the fraction of the worlds with intelligent life which ultimately produce communicative technologies. Mathematically, this is expressed as:

$$f_{civ} = f_{hab}\, f_{life}\, f_{int}\, f_{tech} = (0.0020)\,(1)\,(1)\,(0.5) = 0.0010$$

Only one out of a thousand stars can be expected to produce a communicative civilization.

THREE TYPES OF CIVILIZATIONS

Even though technological civilizations should appear at great frequency throughout the galaxy, there are many possible alternative technologies that could appear. The Soviet physicist Nikolai Kardashev has classified hypothetical extraterrestrial communicative societies into three chategores: *Type I*, *Type II*, and *Type III*. They differ from each other in the level of sophistication reached by their technologies, particularly in the amount of energy that is accessible for the performance of useful work. Type I civilizations are societies that are at the same level as present-day Earth or perhaps slightly above. They produce a total average power output of the order of 4×10^{12} watts (4 terawatts), about the current level of power generated by all the nations on Earth. Type II civilizations are those capable of harnessing an average power output equivalent to the luminosity of a star, which is roughly 4×10^{26} watts. Type III civilizations are able to command energies of the order of the luminosity of the entire galaxy, or 4×10^{37} watts.

Type I extraterrestrial civilizations can be pictured as somewhat similar in overall form to present-day human society. Such civilizations have achieved control over the environment on their home worlds and may even have been able to modify the planetary surface itself to meet their needs. Their energy is provided by fossil fuels, by nuclear power, by solar collectors, or most likely by a combination of all three. They are probably capable of interplanetary space travel, but they have not yet been able to establish any significant presence on other worlds. Type I civilizations are incapable of large-scale interstellar space flight, but they might be able to communicate with beings from other stars by an exchange of radio signals. Like the human race, other Type I societies may be faced with grave problems posed by their rapid technological growth. These provide an ever-present and growing threat to their survival.

Type II civilizations are envisaged as far more advanced societies which are able to harness the entire energy output of their home star. They are able to perform engineering feats on the scale of an entire solar system, whereas we humans have not yet been able to alter even the surface of our own planet to any significant degree. The environments of barren and hostile planets may have been purposefully changed to make them suitable for habitation. They may have moved whole worlds from one orbit to another to suit their desires and may even have been able to change the structure and luminosity of their central star to delay its departure from the main sequence. Their primary energy source is undoubtedly the solar energy emitted by the central star. Some Type II civilizations may have been able to launch exploratory missions to other nearby star systems in search of new worlds to explore. Advanced Type II civilizations may have complete genetic control over their populations, circumventing natural selection to produce individuals bred to precise specifications. Individuals can probably live indefinitely, death coming only by accident or by choice. Advanced computer technology may have produced a machine intelligence far superior to the organic intelligence which created it. All mature Type II civilizations may be entirely computer-based, with superintelligent machines replacing the carbon-based life forms that created them.

There may be many highly advanced Type II societies in the galaxy, descended from those Type I civilizations which managed to survive technological adolescence. These civilizations may be politically and economically stable for thousands or even millions of years, having learned how to live with the fruits of their technological progress without being swamped by the side effects. These societies will have access to the virtually unlimited energy of the ever-shining star near which they live. Their accomplishments must be brilliant to behold. Perhaps we can contact them. There is much that they can teach us.

One possible form that a Type II civilization might assume has been suggested by the American physicist Freeman J. Dyson, of Princeton University. He pictures a Type II civilization as one which has been able to construct a huge spherical shell of radius 1 to 2 AU around its sun, so that all the solar energy produced is trapped. This shell might be constructed out of the material contained within Jovian-sized planets and could be as thin as two or three meters. Solar collectors would be mounted on the interior surface of the shell to gather the sunlight which would power all the industrial and technological activity of the civilization inside. A "Dyson sphere" (as such a structure has come to be known) could not be seen visually

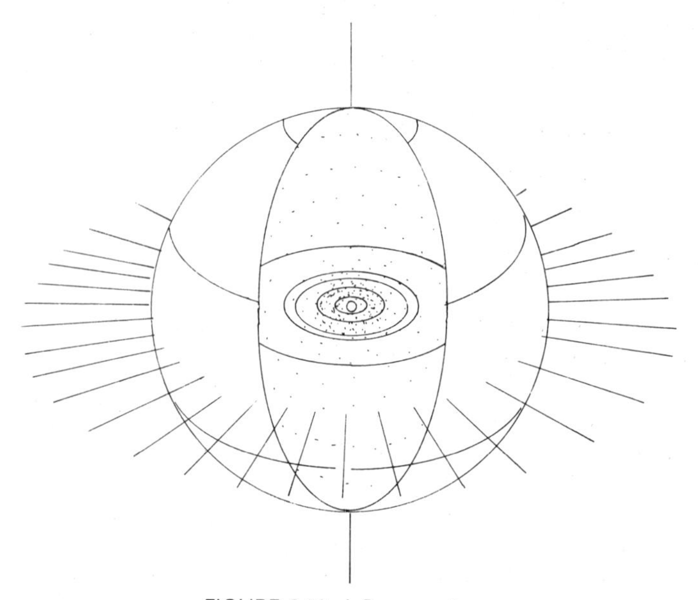

FIGURE 3.10. A Dyson sphere.

from the outside, as all the light emitted by the star would be blocked. It could nevertheless still be detected, as the outside surface of the sphere must radiate just as copiously as the star inside. This radiation would be in the far infrared region of the spectrum (at a wavelength of 10 microns), rather than in the visible. Some doubts have been expressed about the technical feasibility of such a huge object as a Dyson sphere. To begin with, there is probably not enough solid material in our own solar system to construct such an object (Jupiter is largely hydrogen), and even if there were the sphere probably could not remain mechanically stable. Nevertheless, Dyson believes so strongly in the possibility of such spheres that he would regard any conclusive proof of their nonexistence as evidence for the absence of any other technological civilizations in the galaxy.

The American physicist Gerard O'Neill (also of Princeton University) has suggested a more likely form that a Type II civilization might take, one which the human race itself might eventually produce. He has proposed that NASA consider the construction of large artificial colonies in space near the Earth. These would be kilometer-sized cylinders with thousands of people living on the inside. Such structures could be rather light, since they would not have to support any weight. Solar energy is reflected into the cylinder by a system of mirrors. An artificial gravity is provided by spinning the cylinder about its axis. The atmosphere is maintained by establishing a miniature Earth-type oxygen–carbon dioxide ecological

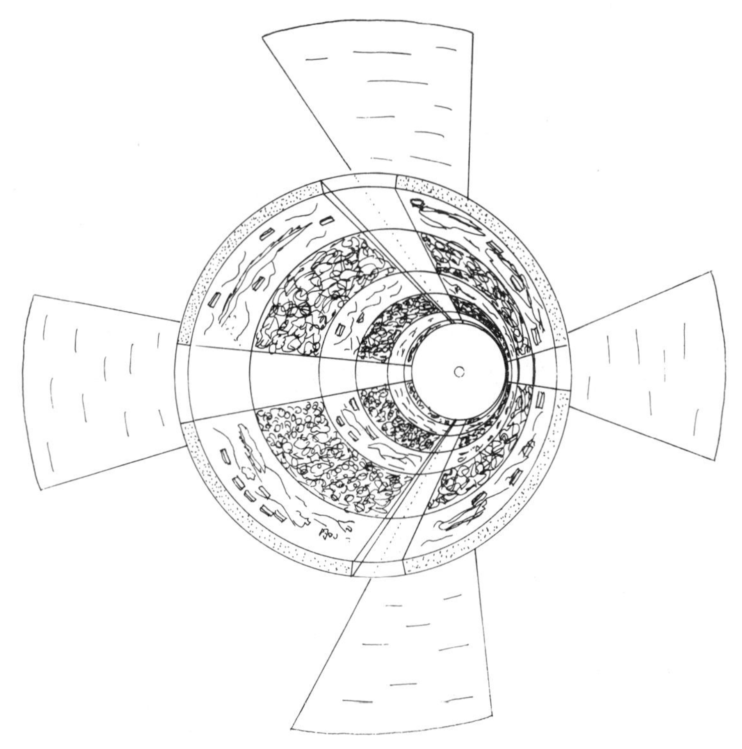

FIGURE 3.11. Interior of a future space colony.

cycle inside. The materials required for these colonies can be provided by metals mined on the Moon and fired aloft by a series of magnetically propelled launchers. At a later time, the asteroid belt could be exploited. The first few colonies would be built near the Earth, but in succeeding years more elaborate space colonies could range further out into space. In the next few centuries, there may be a rapidly growing "cloud" of space colonies surrounding the Sun. The space-living population could reach a billion individuals by the year 2700. Most of our descendants may live in outer space, the Earth having been set aside as a sort of park or wildlife preserve. O'Neill regards the establishment of artificial space colonies as the only realistic answer to the coming crises of resource exhaustion, population overcrowding, and environmental pollution that threaten human technological civilization.

What would a Type III society look like? This is a society which has the energy equivalent of an entire galaxy at its disposal. Such a society could have originated from an ancient Type II civilization driven by population pressures, depleting resources, the end of its star's main-sequence lifetime, or even the threat of a nearby supernova to expand outward into the galaxy. A society capable of commanding such enormous resources probably obtains its primary energy source from the direct conversion of massive amounts of matter entirely into energy. Alternatively, they could tap the luminosity of every single star in the galaxy. They may be able to extract energy in virtually unlimited amounts by allowing matter to collapse into massive black holes. With access to such vast energy resources, engineering feats on the scale of the width of the entire galaxy would now be possible. Type III civilizations may be able to alter the structure of entire star clusters or even the galactic nucleus itself to suit their needs. They may be able to control the process of star creation itself, manufacturing stellar objects more to their liking. Type III civilizations are undoubtedly capable of making large-scale interstellar voyages at near-light velocities. The durability of such advanced civilizations is no longer limited by the lives of the stars; when a life-supporting star dies, the society surrounding it can simply move to another. The individuals in a Type III society may be superintelligent machines, originally constructed by carbon-based life forms long since perished. These machines may be linked together in a vast communicative network that spans the entire breadth of the galaxy, effectively creating a single superbeing capable of virtually unlimited feats of intellect. The works of a Type III civilization would seem to us as nothing short of magic. They would appear to us as gods.

One hypothetical Type III civilization that is a favorite of science-fiction writers is the galactic empire. It is usually visualized as similar to the Roman Empire, with millions of worlds living under the rule of some sort of central authority. Huge starships may be plying the spaceways of the galaxy, carrying the commerce and trade of a million worlds back and forth between the stars. Interstellar conflict may be an unhappy reality, with different worlds and civilizations in constant struggle with each other for galactic domination. Galaxy-wide warfare could be taking place at this very moment, having lasted for a billion years or more with no end in sight. Horrendous weapons may be in use, capable of wiping out entire solar systems at a single blow or causing stars to drift prematurely off the main sequence. The empire may be indescribably vicious and cruel, decimating whole worlds and exterminating entire races to bring the galaxy under its control. If such a cruel empire actually exists, Earth has been extraordinarily fortunate in so far failing to attract its attention.

Could there actually be a Type III society currently present in our galaxy? It does seem rather unlikely that a technology operating on so gigantic a scale could actually exist without our being aware of it. Evidence of a highly advanced Type III civilization may actually permeate our entire surroundings, but it is not recognized because it lies entirely outside the range of our observations, much as television signals are invisible and unintelligible to ants. Nevertheless, several scientists have suggested that serious searches be made for signs of such massive intelligent activity. One place to look may be the center of the galaxy. The galactic nucleus is an intense source of X-ray and radio emissions, and there are several odd-looking infrared sources located there. Something extremely interesting is going on. Such phenomena could be produced by the "wasted energy" emitted by a Type III supercivilization at the galactic center, although alternative natural explanations (such as a massive black hole) must of course also be considered. Type III civilizations could also be present in other galaxies, and several Soviet scientists have suggested that a search be made of these galaxies for evidence of intelligence. Particular attention should be paid to galaxies which have abnormally luminous nuclei, anomalously broad emission lines in their optical spectra, spectra with nonthermal continuua, strong radio emissions, rapid variability, or unstable jets or tails. It will, of course, be difficult to distinguish between galactic abnormalities due solely to natural causes and those that result from the activities of superintelligent beings.

In 1965, Soviet radio astronomers announced that they had spotted two anomalous radio sources, designated CTA 21 and CTA 102, that might be Type III civilizations. Their angular dimensions were quite small (less than 20 seconds of arc), and their radio emission spectra had a shape matched by no known natural radio source in the universe. In addition, radio signals from CTA 102 showed rhythmic fluctuations. This announcement caused a worldwide sensation, but it proved to be premature. Later, more detailed analyses seemed to

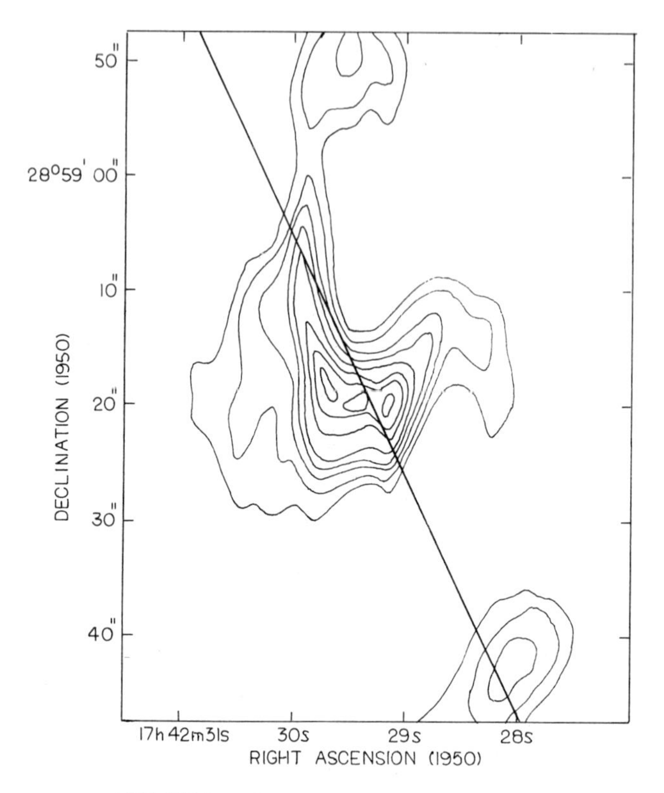

FIGURE 3.12. Radio map of the central arc minute of the Galaxy made by the Very Large Array at a wavelength of 6 centimeters. The diagonal line represents the plane of the galactic disk. From Brown, Johnston, and Lo (1981). National Radio Astronomy Observatory.

indicate that CTA 21 and CTA 102 were probably quasars rather than supercivilizations. Quasars are intensely luminous objects that also produce strong radio emission. All quasars appear to be fleeing from the Sun at enormous velocities, some almost as fast as the speed of light. They are believed to be the most distant objects in the universe, being many billions of light-years from our solar system. Since they are so far away, they must have luminosities exceeding that of entire galaxies to be seen at all. However, they must be physically rather small (200 to 5000 AU across), since they appear to be point sources of light. The entire energy of a galaxy is packed into a space not much larger than a solar system! Astronomers have only the vaguest idea of what sort of an object a quasar really is; the most commonly suggested notion is that they are very early phases in the evolution of galaxies. The astronomical amounts of energy streaming outward from quasars may be produced by supermassive black holes lurking in the centers of such objects. Perhaps our own galaxy went through a

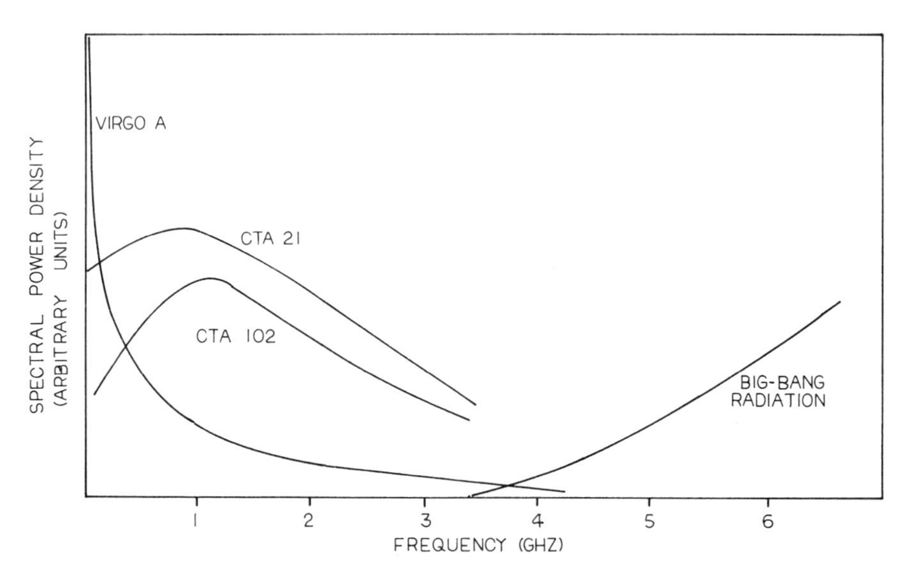

FIGURE 3.13. Spectrum of radio emissions from CTA-21 and CTA-102. From N. S. Kardashev (1964). © 1964, American Institute of Physics. All rights reserved.

quasar phase many billions of years ago, the hypothetical inert black hole at the galactic nucleus being the only remnant of this early violent period. Whatever quasars ultimately turn out to be, most astronomers feel that it is highly unlikely that they will be proven to have an intelligent origin.

THE BIRTHS AND DEATHS OF TECHNOLOGICAL CIVILIZATIONS

In the previous section, it was proposed that only one out of a thousand of the stars in the sky will ever produce indigenous technological societies. Nevertheless, since there are 200 billion stars in the galaxy there could be as many as 200 million places where technological societies might be found. However, it is exceedingly unlikely that this many technological civilizations will actually be present at any one time in the galaxy. Extraterrestrial civilizations will not all arise to communicative status at exactly the same time, and none will last forever. A given habitable planet may be perfectly capable of supporting an advanced technological civilization, but such a society may not yet have appeared or else may have perished a long time ago.

The number of intelligent societies present at any given time in the galaxy will depend on the rate at which new societies appear as well as on the rate at which the old ones die out. An estimate of the rate at which new civilizations appear in the galaxy can be obtained by multiplying the rate at which new stars are created (R^*) by the fraction of those new stars which will eventually produce communicative civilizations (f_{civ}):

$$R_{civ} = R^* f_{civ} \text{ civilizations/year} \tag{3-1}$$

where $f_{civ} = 0.0010$. The galaxy is thought to have formed about 15 billion years ago. Since the total number of stars in the galaxy is approximately 200 billion, the average rate of new star creation over the entire lifetime of the galaxy is approximately $R^* = 15$ per year. This gives $R_{civ} = 0.015$ civilizations/year. A new technological civilization should appear in the galaxy about once every 70 years.

No terrestrial civilization has lasted forever, and it is just as likely that extraterrestrial societies also have finite lifetimes. After all, Earth has been in the communicative phase for only a few decades at best, and our own survival is still in some doubt. Extraterrestrial societies are likely to be faced with similar threats and could meet their doom from a variety of causes. Some possibilities are enumerated below.

Nuclear War

The continued existence of nuclear weapons threatens the life of every human being on Earth. A massive nuclear exchange between the superpowers could end all organized society or even destroy all life on the planet. Published data indicate that the nuclear arsenals of the United States and the USSR contain a total explosive power of 26,000 megatons. This is equivalent to 7 tons of TNT being given to every man, woman, and child on Earth. The World Health Organization has estimated that a war between the United States and the USSR involving the use of most of the nuclear weapons in their arsenals would kill 1.1 billion human beings outright, with another 1.1 billion dying shortly thereafter of burns, disease, or radiation sickness.

The initial consequences of a nuclear war are not nearly so horrible as what would come afterward. The nuclear explosions and resultant fires would lift a tremendous amount of dust and smoke up into the stratosphere. This material would eventually spread out to cover the entire world. This global cloud would reduce the sunlight at ground level to only a few percent of normal. Surface temperatures in most inland regions could drop to −25 degrees Celsius, even for a summer war. This "nuclear winter" could last for months, killing crops and farm animals and disrupting the cycle of green plant photosynthesis. Most of the dust and smoke would settle back to earth, and temperatures would eventually return to normal. However, many of these particles would be radioactive. The incidence of radiation sickness and cancer would rise. Nuclear explosions would chemically oxidize lots of atmospheric nitrogen. These oxides would react with the ozone in the stratosphere, disrupting Earth's screen against ultraviolet light. The incidence of skin cancer would increase dramatically, but hardest hit may be the aquatic microorganisms which are the beginning link in the Earth's food chain. The ecology of the entire Earth might collapse, bringing an end to all life. A full thermonuclear exchange would be as great a disaster for Earth as the late Cretaceous event which destroyed the dinosaurs.

Every year that passes without effective arms control brings the world closer to the final catastrophe. The United States was the first to possess nuclear weapons. Then the USSR found it necessary to acquire them. Then Great Britain, France, China, and finally India developed nuclear arsenals. Israel and South Africa are persistently rumored to have a few nuclear weapons. Many other nations have the capability for nuclear weapon manufacture; all that is needed is the will to build them. Some terrorist organizations could gain access to nuclear bombs in the near future. A minor dispute, an act of insane desperation, or even an accidental explosion could be the trigger of a worldwide conflagration that destroys human society forever.

Is such a horrible fate inevitable? Humankind has inherited many traits from its reptilian and mammalian ancestors which were invaluable survival aids in the past: aggression, territoriality, fear of strangers, ritual, and blind obedience to leaders. Now that weapons of mass destruction are available, the very traits essential for our ancestors' survival threaten us with destruction. Humans may be preprogrammed by their genetic heritage to self-destruct once they have acquired the capability to do so. It need not be so. Humans have also inherited other traits: compassion, sociability, cooperativeness, love for our children, and innate intelligence. Perhaps by use of the higher centers of our brains we can overcome those baser instincts that drive us toward self-destruction.

Extraterrestrial societies are likely to be faced with similar threats once they emerge into the communicative phase. They too will have inherited dangerous psychological traits that place their survival in jeopardy once they have access to nuclear weapons. It is likely that every civilization in the galaxy will be faced with the challenge of the technology of mass destruction. Some societies may have mastered the challenge and survived to build brilliant civilizations, whereas countless others may have chosen death.

Overpopulation

Modern times have seen a particularly dramatic rise in the number of people in the world. At the time that Columbus discovered America the global population was about a quarter of a billion people. By 1650 the population had doubled, and by 1950 the population had reached 2.5 billion. The global population now stands at 4.3 billion. In 1798 the English economist Thomas Malthus published his *Essay on Population,* in which he asserted that human populations inevitably tend to grow so rapidly that they ultimately exceed the resources available to feed, house, and clothe them. When this happens, war, famine, pestilence, and misery are the inevitable result. Malthus argued that the rate of population increase should be proportional to the number of individuals presently living. Mathematically, this is expressed as:

$$dP/dt = aP \tag{3-2}$$

where P is the human population and a is the rate of population increase per person. Currently, $a = 0.02$ per year. Solving Equation 3-3 for P gives:

$$P = P_o \exp(at) \tag{3-3}$$

where P_o is the initial population. This is known as *exponential growth.* According to this law, the human population should double every 25 years; there should be 6.5 billion people by the year 2000. Malthus proposed that the available food and energy supply grows only linearly, and the human race would inevitably become too large to support itself. Nearly two hundred years have passed since Malthus's essay, and no such worldwide catastrophe has occurred. Fortunately, the growth in energy and food supplies has so far kept pace with human population growth. However, the rapid population increases that will take place by the end of this century and the beginning of the next will severely tax the Earth's resources and produce potential sources of conflict as one nation or group tries to expand its resources and territory at the expense of another.

Other technological civilizations may also be faced with the problems produced by a rapidly expanding population. The very success of technology and industrialism may ultimately produce a population too large to be supported. Unrestrained population growth in a given society may permit catastrophic instabilities to develop in which massive warfare, starvation, or pestilence is the inevitable result. Such a disaster may result in the end of that particular civilization as a thriving technological society. A collapse of this magnitude may be irreversible, making the unfortunate civilization forever incapable of making further technological progress and incapable of establishing any contact with fellow intelligent beings around other stars. Even if such a disaster is avoided, the effort required for the maintenance of an excessively large population may require all of the energies of a particular society, leaving no room for the exploration of uncharted worlds or communication with other intelligent beings.

Exhaustion of Resources

In recent years, most industrialized nations have become painfully aware of the finite nature of the world's resources. Energy in particular has suddenly become scarce and expensive.

Over the past century America's energy usage has been growing at an exponential rate, with a doubling of energy production every 17 years. This growth has been sustained by cheap and abundant energy resources. Unfortunately, this exponential growth in energy usage cannot continue for much longer. The rate of coal production is projected to peak between the years 2100 and 2150 and will fall off rapidly thereafter as the accessible coal becomes depleted and new coal becomes harder to find and more expensive to extract. The world oil picture is even bleaker. Oil production will peak about the year 2000 and will begin to decline after that as new oil becomes harder to find and more difficult to get out of the ground. On human historical time scales, the era of fossil fuel usage will be incredibly short.

The world economy is so utterly dependent on the availability of cheap fossil fuels that it is difficult to contemplate what will happen when they become scarce. Many of the world's nations are contemplating an extensive development of nuclear energy as one answer to the coming fossil-fuel crisis. In nuclear power, the energy is provided by the fission of uranium-235 nuclei by reaction with slow neutrons. Unfortunately, only 0.7 percent of all uranium consists of this isotope, the bulk being made up of the nonfissionable uranium-238. Uranium ore must be artificially enriched (to about 2.5 to 5 percent) in U-235 before it is of any use as fuel. The mining, processing, and enrichment of uranium ores are extremely expensive operations. In addition, the large-scale consumption of uranium produces a lot of radioactive waste products, some of which can be dangerous for thousands of years afterward. Nuclear power plants are extremely complex and delicate, requiring extensive safeguards and precautions to prevent even the slightest accident from spreading radioactive poisons for miles around. The ready availability of fissionable materials creates the ever-present danger that some may be illegally diverted into nuclear weapons manufacture by desperate underdeveloped nations, by terrorist groups, or even by deranged individuals. Even if most of these problems are satisfactorily resolved, there is still probably only enough high-grade uranium ore present in the United States to last until the end of the century. After that, nuclear power using uranium as a fuel may become prohibitively expensive.

Recently, attention has been paid to the development of breeder reactors which could make far more efficient use of uranium. In breeder reactors nonfissionable U-238 is converted into fissionable plutonium-239, so that virtually every uranium atom present in the ore can ultimately be used as fuel. The breeder reactor actually produces more fuel than it uses; it increases the amount of energy that can be extracted from a given mass of uranium a hundredfold. Furthermore, lower-grade uranium ores can be used. The energy potentially available from uranium ore is at least a few orders of magnitude greater than that available from all fossil fuels combined. Unfortunately, the problems involved in the use of breeder reactors are at least an order of magnitude more severe than those involved in the use of more conventional reactors. The fission reaction within the core is much more difficult to control; the slightest miscalculation or failure could lead to a catastrophic explosion that could spread radioactive poison over an entire state. Plutonium is one of the most toxic substances known to man and is also one of the prime ingredients in nuclear weapons. Even if the problems involved in breeder technology are satisfactorily resolved, the total uranium available in the ground may only be enough to sustain current energy growth for another century or so.

The ultimate solution to the energy problem must lie in the development of entirely new sources of energy. Several nations are investing heavily in research on thermonuclear-fusion power generation. This is the same source of energy that is responsible for the Sun's heat. There is enough thermonuclear fuel in a cubic kilometer of ocean water to provide all the energy needs of the United States for the next fifty years. Fusion-energy plants promise far fewer safety or environmental hazards than nuclear fission plants. However, before fusion power can become a reality a whole host of technological problems must be solved. Extremely high temperatures must be reached before nuclei can overcome their mutual electrostatic repulsion and fuse. The attainment of such high temperatures for a sustained period of time has so far proven to be an impossible task, and no practical controlled and long-lasting fusion reaction has yet been achieved, even in the laboratory. Until these problems are solved, fusion power will remain only a vague hope for the distant future.

Other workers are pushing hard for the development of solar power. Enough solar energy falls on a half of a percent of the United States to provide all of our projected energy needs for the near future. Up to the present, comparatively little effort has been invested in solar power research. Sunlight is so diffuse and intermittent that collection and storage systems have been difficult to develop. The cost of solar power generation currently so far exceeds that of fossil fuel or nuclear power that it is rarely used, except for certain specialized applications, such as power for spacecraft. As fossil fuels become more expensive or as the dangers involved in nuclear power become more readily apparent, solar power could become an attractive alternative.

The rapidly approaching fuel crisis will tax the creativity and ingenuity of the entire human race. Some means must be found of providing abundant, cheap, and safe energy if we are to sustain our growth. Unless we do, most of the human race will be faced with ever-increasing shortages, high inflation rates, and a steadily eroding standard of living. Energy shortages could eventually become so severe that they bring about a complete collapse of industrial society and a return to the days when most work was performed by human and animal muscle. Unless we plan wisely, our present, highly industrialized technological society will be only a passing phase in human history, after which we slowly but inevitably drift back into the primitive hunter-gatherer lifestyle from which we arose. We will cease to exist as a communicative society with the ability to travel in outer space and communicate with other intelligent races in the stars.

All technological civilizations, no matter how sophisticated, will ultimately have to face the problems provided by the finiteness of their material and energy resources. The stars may give birth to many intelligent races that arise to create impressive technological societies, only to squander all of their precious resources in a few hundred years of wanton abandon. Such profligate societies will inevitably fall back into their original, primitive, pretechnological state, fated forever to remain out of touch with the rest of the galaxy.

Pollution

Environmental pollution has been one of the more unpleasant side effects of human technology. In particular, the burning of fossil fuels produces a number of noxious chemicals that can be severe health hazards.

The combustion of coal invariably produces sulfur dioxide (SO_2) as a by-product. Sulfur dioxide in the air is responsible for a variety of respiratory ailments and has been implicated in a number of air-pollution disasters, such as the one which struck London in 1952. When SO_2 becomes mixed with water mist in the air, a strongly acidic rain is produced which kills living tissue, destroys clothing, and deteriorates building materials. The burning of coal also produces soot particles. Many of these are potent carcinogens, and a conversion back to a coal-fired economy could produce an increase in the incidence of cancer.

Most of the oil consumed in the United States goes to the automobile. American automobiles annually release 150 million tons of carbon monoxide, 40 million tons of hydrocarbons, and 24 million tons of nitrous oxides, plus hundreds of thousands of tons of lead compounds. Carbon monoxide is colorless and odorless, but is extremely toxic. A concentration of only one part per hundred in the air can be fatal. Nitrogen oxides (collectively labeled NO_x) are produced in any combustion process that takes place in air because of the presence of atmospheric nitrogen. NO_2 in high enough concentrations can produce acute toxic effects, but in lower amounts it can contribute to chronic respiratory ailments. It is one of the key substances entering a chain of chemical reactions that produces "smog," a common occurrence in such cities as Los Angeles where automobile usage is heavy. The lead compounds poured into the air are all highly poisonous. Ozone (O_3) is produced when the hydrocarbons in automobile exhaust react with the oxides of nitrogen in the presence of sunlight. When the concentration of ozone exceeds 0.1 part per million, toxic and irritating effects take place, damage to vegetation occurs, and organic materials such as rubber become degraded.

The burning of any fossil fuel, whether it be coal, oil, or natural gas, invariably produces carbon dioxide. Carbon dioxide is a normal component of the atmosphere and is an essential part of the carbon cycle in the biosphere. Industrial activity during the last century has produced a steady increase in atmospheric CO_2 concentration. It is estimated that the average CO_2 concentration has risen from 280 ppm to 315 ppm in the last hundred years. The current rate of increase in CO_2 content is about 0.7 ppm per year. If this trend continues, in the year 2025 atmospheric carbon dioxide will be twice the level it was during the early nineteenth century. Could this excess CO_2 adversely affect the Earth's climate? Carbon dioxide in the atmosphere acts to trap heat near the surface. If the carbon dioxide concentration gets too high, there could be an alarming rise in the average surface temperature over the entire Earth. The polar icecaps would begin to melt, flooding many coastal areas. Estimates indicate that a doubling of the present CO_2 concentration would increase the average global temperature by 2 to 3 degrees Celsius. Were the CO_2 concentration to increase by two or three orders of magnitude over the current level, the average surface temperature could get 20 to 30 degrees warmer. The Earth could be forced into a "runaway greenhouse" effect in which the oceans all boil away and the surface conditions become much like those on Venus. We do not yet know whether such a catastrophe could be produced by excessive fossil-fuel burning on Earth because we do not yet know enough about how the world's ecosystems respond to rapid changes in carbon dioxide concentrations. The oceans are a major sink for carbon dioxide, as are the solid carbonates in the Earth's crust. The total animal and vegetable biomass is also a factor. The warming effects of carbon dioxide may be offset by the increased light-scattering in the upper atmosphere produced by particulate pollutants. The human race may be lucky or may have already sealed its doom. Only time will tell.

All energy consumption, whether it is generated by the burning of fossil fuels, nuclear fission, hydrogen fusion, or solar power, ultimately produces waste heat as the final product. How much larger could Earth's rate of energy production become before its surface temperature would begin to increase by a significant amount? An increase in the average global temperature of only a single Celsius degree would produce severe changes in the ocean levels, rainfall patterns, and cloud cover. Such a temperature increase could be produced if the average global rate of energy generation rose to a level as high as 1 or 2 percent of the total incident solar energy. This would be equivalent to the Sun suddenly becoming 1 or 2 percent brighter. The average global rate of energy output required to accomplish this feat is 2000 to 4000 terawatts, which is only a few hundred times the current rate. If the rate of energy production continues in the future to double once every 17 years, as it has in the past, scarcely a century from now humanity will be faced with rapidly rising global temperatures resulting from the waste heat produced by its own energy consumption. Growth would have to cease lest we smother ourselves to death.

Environmental pollution seems to be an unavoidable consequence of technogical progress. All societies that are dependent upon the consumption of large amounts of energy will be faced with a rapid buildup of unpleasant and dangerous waste substances. If allowed to proceed unchecked, environmental pollution could force an abrupt end to technological society. There may be many intelligent civilizations in the stars which were forced to restrict the scope and sophistication of their technological activity or even abandon it altogether rather than be faced with a slow death by poisoning. Others may have ignored the problems posed by unchecked pollution for so long that they were forced to divert their entire efforts to cleaning up the mess. Some civilizations may have recklessly continued to increase their energy usage at an exponential rate until they were smothered to death in their own waste heat. There may even be many unfortunate worlds that were too late in recognizing the danger of the "runaway greenhouse" effect produced by unrestrained fossil fuel consumption and were subsequently turned into lifeless Venus-like ovens.

Genetic Deterioration

Throughout the entire duration of evolutionary history, life had advanced in adaptability and intelligence through the complementary processes of mutation and natural selection. Random mutations produce changes in the offspring of a given species. If the mutation is unfavorable, the offspring will generally die before it has a chance to reproduce. If the mutation is favorable, the new organism will live and prosper, passing its traits on to successive generations. It is in this manner that evolution has managed to produce so many different types of creatures, each well adapted to its own particular environment.

With the advent of technology, the picture has changed. Advances in medicine allow not only the fittest to survive, but almost everyone else too. Weak or sickly individuals tended to die at a young age in the past, whereas medicine now makes it possible for them to live long enough to produce offspring. As the treatment of hereditary disease increases in effectiveness, more and more people who have them are able to live and pass their diseases on to successive generations. The advent of modern technology has added myriads of potential mutagens to the environment, such as radioactive fallout, leaks from nuclear power plants, soot from smokestacks, by-products of chemical manufacturing, and a whole host of others. As these substances increase in concentration, more and more mutations may be produced in the population at large. The vast majority of these mutations are likely to be unfavorable, and in the absence of natural selection they could be passed on to succeeding generations. There could be a slow but steady deterioration in the physical and mental capabilities of the entire human race over the next few hundred years.

Another set of dangers has been raised by the recent developments in biochemistry that may soon make it possible to alter the human genetic structure at will. By the use of genetic engineering techniques, the genes that code for desirable characteristics can be promoted, whereas those that produce undesirable characteristics can be eliminated. We may ultimately be capable of creating entirely new types of human beings to any set of specifications we choose. The advent of this new technology naturally raises the question as to what authority will decide what traits of intelligence, personality, or physical appearance are to be favored and which ones are to be suppressed. In Aldous Huxley's *Brave New World,* such decisions were made by the state, and individuals were "manufactured" according to an overall plan to fit into a previously established class structure. A static society lacking in essential creativity was the result. Genetic engineering technology could be seized by racist fanatics who believe that they and they alone know how to create the "superman," much in the same spirit with which the German Nazis dabbled in genetic studies during the 1930s and 1940s. We know so little about what intellectual, physical, and personality traits are important in the overall scheme of human existence that we embark upon such activities at our peril.

Genetic deterioration may be a real threat for those extraterrestrial societies that are fortunate enough to avoid destruction by nuclear war, pollution, overpopulation, or exhaustion of their natural resources. The effects of environmental pollution combined with the cessation of natural selection may have produced many civilizations with populations too weak or too stupid to sustain any sort of technological society at all. Some civilizations may have hastily embarked upon ill-advised programs of massive genetic engineering with disastrous consequences. They guessed completely wrong about the traits most needed for survival and technological growth, the new race quickly dying out from unforeseen diseases or else being physically or emotionally incapable of any sort of technological life at all.

Overstabilization

The Earth's population and energy output cannot continue to grow much longer at the present exponential rate without eventually producing a worldwide crisis when pollution, overcrowding, and energy shortages reach the critical point. This could happen as early as the first decades of the twenty-first century unless steps are taken now to prevent it. One possible alternative is for the nations of the world all to agree to halt the growth of their economies. This "no-growth" option has been forcefully put forth by the "Club of Rome," an unofficial group of economists and social scientists concerned about the coming ecological and demographical crisis threatened by unrestrained growth. Steady economic and technological growth has been a part of our lives for so long that it is difficult to imagine what the world would be like without it. Presumably, there would be a heavy emphasis on recycling of scarce materials, a large effort to reduce the numbers of polluting industries, and a search for alternative energy sources. It may prove necessary for the size of the population to be rigorously controlled by the state; it may be illegal to have children without some sort of special license. If such a program is to be successful, we must first abandon the assumptions that economic growth is automatically a good thing and that we must continue to expand our use of energy and materials in order to avoid stagnation. Perhaps we humans can adopt a more relaxed style of life that emphasizes things, such as knowledge acquisition, information manipulation, philosophical contemplation, or personal development, that do not require such large energy and material inputs.

There are dangers in adopting a "no-growth" option. The forcible elimination of growth may come at the cost of establishing an overly static society in which scientific progress and intellectual curiosity have been destroyed. Scientific knowledge and economic growth have long been parallel processes, and the forcible elimination of one could lead to the loss of the other. Scarce energy and material resources may have to be used so carefully that there will be none left over for such "wasteful" activities as space exploration, particle physics, or astronomy, which do not promise an immediate return. One of the benefits that economic growth has brought to most people has been a steady expansion in the number and variety of choices that they can make in living their lives. In a no-growth society these options will be sharply curtailed, and most people will be forced to live dull and marginal existences. That zest for life which is so necessary for intellectual creativity may be lost. The curiosity about the world that has characterized human beings for so long may be dampened. We may turn inward, abandoning the exploration of the universe that we had begun so confidently such a few short years before.

Other civilizations in the galaxy may have solved the problems of pollution, warfare, overcrowding, or energy shortages at the cost of establishing overly static societies where technological and intellectual progress have ceased. There may be many worlds that chose to halt their development at the Type I level. They will forever be chained to their home planets because they lost their will before they could establish any significant presence in outer space. They will slowly but inevitably slide backwards into primitive, pretechnological lifestyles from which there is no hope of recovery. Over the long term such confined societies may undergo a slow but steady intellectual stagnation. They could even become quite xenophobic, resenting any intrusions from the outside. Such a fate could await Earth if it abandons its expansion outward into the solar system.

THE GREEN BANK FORMULA

Let the average lifetime of technological civilizations of whatever type (I, II, or III) be designated by τ_{av}. We can interpret τ_{av} as the mean time that extraterrestrial civilizations exist in the communicative phase, during which they are capable of and interested in establishing contact with other intelligences. We of course have no idea of the length of τ_{av}. It could be as short as a few decades or it could be as long as the lifetime of the galaxy itself. If τ_{av} is only a few decades, we can conclude that few if any advanced civilizations in the stars survive the perils of technological adolescence that beset current-day Earth. Virtually all newly emergent communicative civilizations would destroy themselves before they got the chance to establish any sort of interplanetary culture. Earth itself statistically stands very little chance of long-term survival. If τ_{av} ranges from a few thousand to a few million years, then many societies are able to establish highly advanced interplanetary Type II civilizations. If the average lifetime is truly this long, it may mean that the Earth stands at least a fighting chance of surviving the crucial dangers ahead. The human race may have many, many years of fruitful life ahead. If τ_{av} is longer than millions of years, then extremely ancient societies are fairly common in the stars, and many Type III societies could exist in the galaxy.

The rate at which civilizations perish is equal to the number of civilizations currently in the galaxy divided by the average lifetime:

$$D_{civ} = N_{civ}/\tau_{av} \qquad (3\text{–}4)$$

The net rate at which new civilizations appear (or disappear) in the galaxy is equal to the difference between the birth rate (Equation 3–1) and the death rate (Equation 3–4):

$$dN_{civ}/dt = R^*f_{civ} - N_{civ}/\tau_{av} \qquad (3\text{–}5)$$

There must have been a time when there were no technological civilizations present in the galaxy. During the first few billion years of galactic history, Population II stars were the only types of stars in existence. Their planets (if they had any) were too deficient in heavy metals to support any type of life at all, to say nothing of advanced technological societies. However, as the early massive stars exploded and died, the interstellar medium was gradually seeded with the elements that were to prove essential for life and technology. At some unknown time in the distant past the first technological society was born. For a long time there were very few such civilizations in the

TABLE 3.1 Estimate of the Current Number of Extraterrestrial Civilizations and Their Average Separation for a Range of Mean Communicative Lifetimes

Mean Lifetime (years)	*Number of Civilizations*	*Mean Separation (light-years)*
50	1 (us)	—
100	2	12,000
500	10	8800
1000	20	6300
10,000	200	2900
100,000	2000	1400
1,000,000	20,000	630
10,000,000	200,000	320
100,000,000	2,000,000	140
1,000,000,000	20,000,000	50

stars, but their numbers must have steadily increased as the years passed. As newer societies came on line, some of the older ones must have perished. Eventually a *steady state* was reached in which new societies were appearing and old ones were passing out of existence at exactly the same rate. The galaxy is usually assumed to be in this steady-state phase of the evolution of intelligence at present. This idea is derived from the so-called "principle of mediocrity," which proposes that there is nothing special about the location of the Earth in space and that there is nothing unique about the particular instant of time during which the human race has appeared. We are probably average and typical rather than extraordinary or special. If the steady-state model is valid, then the number of communicative societies is constant in time and the quantity dN_{civ}/dt in Equation 3-5 is equal to zero. The number of communicative civilizations currently in the Galaxy is then equal to:

$$N_{civ} = R^* f_{civ} \tau_{av} \tag{3–6}$$

This formula was first expressed by Frank Drake in 1961, and has dominated all discussions of the probabilities for extraterrestrial civilizations ever since. It is sometimes called the "Green Bank" formula, after the location of the radio telescope that carried out the first search for intelligent signals from outer space.

We of course have no knowledge of the mean lifetime of extraterrestrial civilizations, so we have no way of reliably estimating the current number of civilizations in the galaxy. However, we can see some trends if we calculate N_{civ} from Equation 3–6 for a range of assumed lifetimes. This is given in Table 3.1. Several things are apparent. If τ_{av} is only a few decades, the number of communicative civilizations must be very small. We could actually be the only communicative society in the galaxy at this particular moment, although millions of others may have risen and fallen before we appeared. We may be destined to share their fate. We can anticipate large numbers of intelligent civilizations only if mean communicative lifetimes are quite long, of the order of millions of years or more. Only if the probability of long-term survival is quite high can we expect to find others like us in the stars. Our own future is thus intimately related to the question of the existence of technological civilizations around other stars. If we find that we are alone in the galaxy, this fact in itself may mean that our chances for long-term survival are rather slim. If other technological societies are present, perhaps their mere existence means that we too have a future. Even if there are large numbers of intelligent societies, their mean separation is likely to be hundreds or even thousands of light-years. Enormous distances will have to be breached if we are ever to make contact with these creatures.

CHAPTER FOUR

THE ERA OF THE STARSHIP

Human beings have taken their first tentative steps into outer space, hopefully one day to establish a civilization encompassing the whole breadth of the solar system. Unmanned exploratory craft have already flown past all of the planets within the orbit of Uranus. Spacecraft routinely carry astronauts on orbital missions lasting as long as several months. Men have walked upon the Moon. Both the Soviet Union and the United States have begun research aimed at establishing permanent space stations in near-Earth orbit, the first vital step in the colonization of the Solar System. Assuming that we manage to avoid self-destruction, human beings can reasonably expect to be able to explore and colonize the entire solar system within the next few hundred years.

Can human beings realistically expect to travel even farther into outer space, ultimately to other stars? It now seems certain that the human race is the sole technological civilization in the solar system; we must look to other stars in order to find nonhuman, intelligent beings similar to ourselves. What are the prospects of humans being able to send forth space missions to other star systems to seek out these alien civilizations? Even if there actually are large numbers of technological civilizations in the galaxy, such societies can nevertheless be expected to be hundreds or perhaps even thousands of light-years apart. The largest velocities yet attained by spacecraft launched from Earth are of the order of ten kilometers per second. At that rate, a voyage to even the nearest alien civilizations will take several million years.

Does the prospect of such long travel times necessarily make interstellar travel infeasible? It may be possible to transport human beings on long voyages across the galaxy by placing them in some sort of suspended animation in which biochemical processes are slowed down a millionfold. The inhabitants of the ship would be put to "sleep" at the beginning of the voyage and revived at its end, perhaps several million years later. Perhaps interstellar transport can be accomplished by constructing huge starships capable of supporting thousands of generations of people which are born and die during the long voyage. The individuals beginning the voyage would not live to see its end, but their distant descendants would.

The prospect of an interstellar voyage taking thousands or millions of years to complete is not very inviting, at least as far as we humans are concerned. Perhaps future technological advances will make it possible to build spacecraft with appreciably higher velocities, so that the travel times could be brought down to more manageable levels. Is it possible to build spaceships which can travel at velocities that are appreciable fractions of the speed of light?

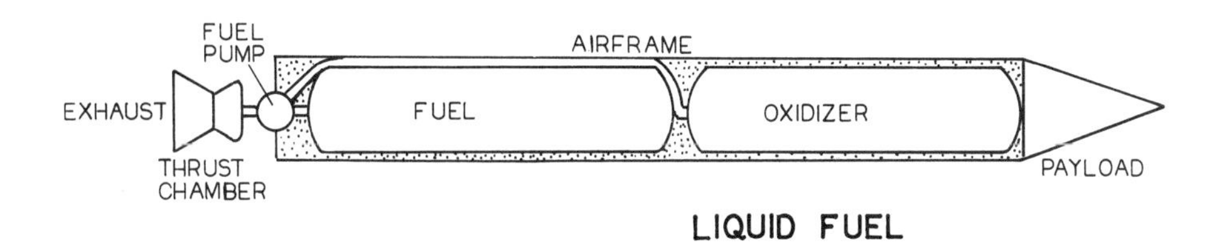

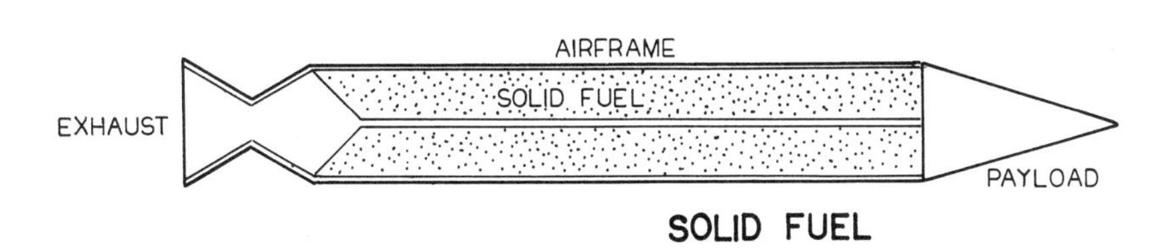

FIGURE 4.1. Schematic view of solid and liquid fuel rockets.

CHEMICAL ROCKETS

All space vehicles constructed on Earth to date have been propelled by the consumption of chemical energy. The means by which this chemical energy is converted into kinetic energy of motion is the *rocket vehicle.* All rockets, of whatever type, consist of four basic components: an airframe, a payload, a propulsion system, and a load of fuel. When a rocket is fired, fuel is admitted to the combusion chamber of the propulsion system and is burned. The highly energetic combustion products thus produced are then expelled from the engine exhaust to provide the thrust.

The largest velocity yet attained by a rocket is 13.9 kilometers per second, first achieved by the Atlas-Centaur rocket that launched Pioneer 10 on its mission to Jupiter. This particular rocket had three stages, the first being a converted Atlas intercontinental missile burning a mixture of liquid oxygen and a chemical propellent similar to kerosene. The second (Centaur) stage burned a mixture of liquid hydrogen and liquid oxygen. The third stage was a specially built solid-fuel rocket engine added to the vehicle to provide the additional velocity increment required to reach a planet as distant as Jupiter.

The final velocity that a rocket can attain after all of its fuel is consumed depends on the exhaust velocity and on the relative fraction of the rocket's original mass which is fuel. In actual practice, most of the takeoff mass of a rocket is fuel, making it possible to achieve a velocity which is appreciably faster than the speed of the reaction products coming out the exhaust.

Is there any prospect that engineering advances in the future will make it possible to construct chemically propelled rockets with appreciably higher velocities? There are only two alternatives open: Either increase the relative fraction of the rocket which is fuel, or else increase the exhaust velocity. Currently available chemical propellents typically give exhaust velocities of the order of 4 kilometers per second. It seems unlikely that any new and exotic chemical fuels will be discovered which will provide exhaust velocities that are appreciably higher. The other option is to increase the fuel/rocket mass ratio. If a rocket is to reach a velocity of 14 km/sec by burning fuel with an exhaust speed of 4 km/sec, 97 percent of the mass of the vehicle must be fuel. In actual engineering practice it is not possible to do nearly as well as this. The total takeoff mass of the Atlas-Centaur booster that launched Pioneer 10 was 147,000 kilograms, but the actual spacecraft which flew past Jupiter had a mass of only 260 kilograms. In order to merely double the final velocity of Pioneer 10, a rocket at least 30 times more massive than the Atlas-Centaur would be required, a booster of the size and weight of the Saturn-5 manned lunar launcher. The situation becomes rapidly worse as even larger velocities are contemplated. To reach a relatively modest 0.1 percent of the speed of light (300 kilometers per second), a fuel-to-rocket mass ratio of 1×10^{32} would be required! There is no prospect at all of radically increasing rocket velocities if we are forced to rely solely on the consumption of chemical energy for propulsion.

NUCLEAR FISSION ROCKETS

Perhaps there are other types of reactions that can produce exhaust products with much greater speeds. What about the enormous amount of energy that is released in nuclear fission reactions? Could it be used to provide rocket thrust? A typical fission reaction is given by the equation:

$$^{235}U + n \longrightarrow {}^{139}Ba + {}^{94}Kr + 3n \qquad (4\text{–}1)$$

Each uranium nucleus that undergoes fission releases about 2 billion electron volts (or 3.2×10^{-10} joules) of energy. About 98 percent of this energy is given to the three neutrons, leaving each fission fragment with an energy of about 20 million electron volts. This translates into a velocity of 6000 kilometers per second. This is a quite respectable velocity, actually a couple of percent of the speed of light and three orders of magnitude faster than the velocities attained by the products of chemical reactions. If the reaction products could be directed into an exhaust, a fission rocket with a mass ratio of 150 could attain a velocity as high as 10 percent of the speed of light. Nuclear energy appears at first sight to offer a real possibility of reaching the nearby stars within a few years instead of many millenia.

There are, of course, a whole host of technical problems which must be overcome before such a propulsion mechanism can become a reality. The storage and handling of such large amounts of nuclear fuel will provide enormous headaches. Nuclear fission reactions invariably produce an intense flux of high-energy neutrons for which it is extremely difficult and expensive to provide adequate shielding. The fission products coming from the exhaust are themselves highly radioactive and will pollute outer space for millions of kilometers around the rocket. The thrust chamber itself would have to be constructed of an as-yet-uninvented super material impervious to the high temperatures and radiation fluxes present during the fission reaction.

During the early 1960s a group of engineers at the General Dynamics Corporation suggested another approach, the *nuclear pulse rocket*. They gave their study the name "Project Orion," and proposed that the uranium fuel be stored in the form of several thousand atomic bombs. These bombs would be ejected one at a time from the rear of the ship and exploded a short distance behind. Some of the energy released by the explosions would be absorbed by a large shield placed at the rear of the ship, producing a regularly spaced series of impulses that would drive the ship forward. To avoid the unpleasant "jerkiness" that such a series of explosions would produce, the impulses would be smoothly transferred to the rest of the ship by immense shock absorbers. Apparently, "Project Orion" was seriously

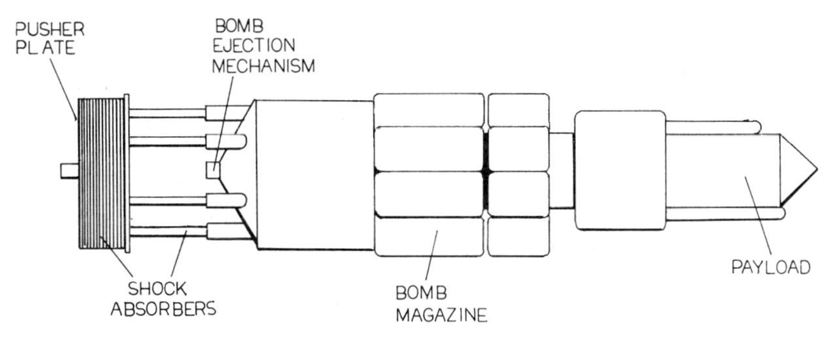

FIGURE 4.2. Orion nuclear pulse starship. From R. L. Forward (1976). Copyright 1976, British Interplanetary Society.

considered by the Defense Department as a possible future means of spacecraft propulsion until the Nuclear Test Ban Treaty of 1963 made atomic explosions in outer space illegal. Freeman J. Dyson has considered some of the details involved in the design of such a ship. He envisages a craft with a total mass of 400 million kilograms, of which 80 percent is 300,000 atomic bombs which are ejected at the rate of one every three seconds from the rear of the ship. This ship could accelerate to a velocity of 1.6 percent of light speed within about ten days, after which half the bombs would be expended. The rest of the bombs would be reserved for deceleration at the end of the trip. Such a ship could reach the Alpha Centauri star system in about 260 years.

Dyson's proposed mission to Alpha Centauri would require approximately a third of a billion kilograms of fissionable material. The manufacture of the atomic bombs would consume a significant fraction of the uranium believed to be left in reserve in the United States. Humanity would have to exhaust almost all of its fission fuel to launch even a single high-velocity starship to a nearby solar system.

THERMONUCLEAR FUSION ROCKETS

The prospect of having to consume the Earth's entire reserve of nuclear fuel in order to launch just one starship does not make one very optimistic about the future of high-velocity interstellar flight. Is there any prospect of finding another source of fuel that cannot be exhausted quite so readily as uranium? Perhaps the thermonuclear fusion process responsible for keeping the Sun alight is a possibility. In a thermonuclear fusion reaction, two lighter nuclei are combined to produce a single heavier one, energy being released in the process. Typical fusion reactions are listed in Table 4.1. Ideally one should

TABLE 4.1 Typical Thermonuclear Reactions

Reactants		Products
$^{2}H + {}^{3}H$	$\longrightarrow$	$^{4}He + n$ (17.59 MeV)
$^{2}H + {}^{2}H$	$\longrightarrow$	$^{3}He + n$ (3.27 MeV)
	$\searrow$	$^{3}H + {}^{1}H$ (4.03 MeV)
$^{2}H + {}^{3}He$	$\longrightarrow$	$^{4}He + {}^{1}H$ (17.81 MeV)

choose a fusion process which has as high an energy yield as possible, which uses only reactants that are readily available in unlimited supply, which uses no materials that are dangerous to process or difficult to handle, and which gives off no hazardous radiations or noxious poisons in the exhaust. It is difficult to find one single reaction which satisfies all of these requirements. The first reaction has a high energy yield, but it has the disadvantage of requiring tritium (^3H), a rare radioactive isotope of hydrogen. In addition, highly energetic neutrons are emitted. Neutrons are a highly penetrating sort of radiation, and propulsion systems producing large numbers of these particles may require massive shielding and extensive cooling equipment. The second reaction requires only deuterium, which happens to be abundant in water and inexpensive to separate. However, the energy yield is relatively low, and neutrons are emitted in one of the two reaction paths. The third reaction shows more promise. The energy yield is high, and no neutrons are emitted. Alpha particles and protons are the sole reaction products; these can be stopped with relative ease by thin layers of material, making the shielding problem much easier to solve. Unfortunately, the reaction requires helium-3, a rare isotope of helium which is only 0.0001 percent abundant on Earth.

In the deuterium–helium-3 fusion reaction the products are a proton and an alpha particle. The energy released in the reaction gives a velocity of 0.03c to the alpha particle and 0.17c to the proton. If these highly energetic reaction products could be directed into an exhaust, spacecraft velocities as high as half the speed of light could in principle be attained, making it possible for a starship to reach nearby stars within the lifetime of the crew and passengers aboard.

Could a spacecraft using this sort of propulsion mechanism actually be constructed in the near future? In the early 1970s the British Interplanetary Society (an organization of spaceflight enthusiasts from many different nations) commissioned a detailed feasibility study for a fusion-powered starship. The study was given the name "Project Daedalus," and the vehicle which ultimately emerged was an unmanned, two-stage, thermonuclear pulse rocket with a total mass of 49 million kilograms. Ninety-three percent of the mass of the craft would be thermonuclear propellent, stored in the form of small pellets of solid deuterium mixed with helium-3. The pellets would be injected one at a time into the center of a large, hemi-

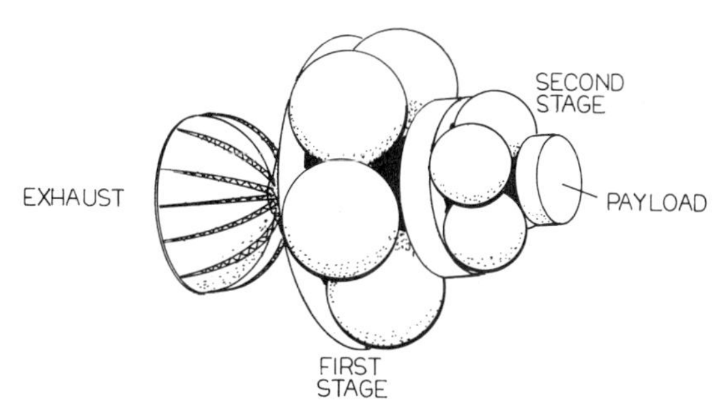

FIGURE 4.3. Daedalus thermonuclear-pulse starship. From Project Daedalus Study Group (1977). Copyright 1977, British Interplanetary Society.

spherical thrust chamber. When each pellet reached the center of the chamber it would be struck by a high-energy electron beam, and the fusion reaction would begin. An intensely hot plasma of electrons, protons, and alpha particles would be created. Intense magnetic fields generated by superconducting coils would confine the plasma and direct it out of the rear of the vehicle to provide the thrust.

The Daedalus team proposed a one-way flypast mission to Barnard's Star (5.9 light-years distant) as the first objective for the thermonuclear pulse starship. The Daedalus ship is projected to be capable of reaching a final velocity of about 13 percent of the speed of light. The boost phase of the mission would last 4 years, and the total trip to the star would take about 50 years. The payload would consist of 450,000 kilograms of instruments and subsidiary mini-probes that would make an intense examination of the star and any planets during the flypast. The encounter phase of the mission would last only a couple of hours.

An erosion shield must be provided at the front of the ship to protect it against the effects of the interstellar gas and dust that will be encountered along the way. A collision with a dust particle no larger than 1×10^{-16} kilograms could permanently damage an unprotected ship traveling as fast as 13 percent of light speed. During the encounter, the ship will almost certainly meet with an even higher density of gas and dust. Even an erosion shield will not be sufficient protection at this time, so a protective screen of fine dust will be squirted ahead of the ship to clear a safe path.

The Daedalus mission to Barnard's Star requires about 27 million kilograms of helium-3. Helium-3 is currently produced in small amounts by the neutron bombardment of lithium inside nuclear reactors. A reactor power level of 200 terawatts would have to be provided over a 20-year time period in order to synthesize the necessary helium. The dedication of such a large power generation capability to the manufacture of starship fuel is quite out of the question, so the Daedalus team suggests that we turn instead to Jupiter for the helium. Seventeen percent of that planet is helium, and it is proposed that a series of separation plants be floated in the upper Jovian atmosphere suspended beneath massive balloons. Automatic shuttle craft would collect the accumulated helium-3 and carry it into outer space for use as starship fuel.

MATTER-ANTIMATTER STARSHIPS

Einstein's special theory of relativity predicts that no material object can ever exceed the speed of light, no matter how much energy is provided. The laws of physics provide the universe with a natural speed limit. This limit makes it physically impossible to travel to a point 4000 light-years distant from Earth in a time period any less than 4000 years. Therefore, at first sight the theory of relativity seems to make the stars forever beyond our reach. However, there is one important qualification that must be added to the time limit discussed above. This limitation applies only for those people who stay at home on Earth, not for those who are making the journey. For the people actually making the trip, it is theoretically possible for them to reach their destination in a much shorter time, one well within their natural life spans. This is the so-called *time dilation* effect, which happens only when a spacecraft is driven to a velocity very close to that of light.

Einstein predicted that the clocks aboard a moving spaceship should appear to an observer at rest on Earth to be moving more slowly than those on the ground. The rate at which time passes for the people aboard the ship is slower than the rate at which time passes back on Earth; all processes, (including physiological aging) appear to be taking place at a much slower rate. This effect is completely negligible during space travel at present speeds because the velocities encountered are much too slow; the slowdown in the rate of time flow is significant only if near-luminous velocities (75 percent or greater) are attained. The closer the velocity of the ship to the speed of light, the greater the effect. Suppose a ship were to leave for a point 4000 light-years away, speeding up at a rate equal to the acceleration of gravity for the first half of the mission, slowing down at the same rate for the other half. The maximum velocity reached at the midpoint of the journey would be 99.999 percent of light speed. The crew and passengers aboard the ship would find that they had reached their destination in only 16 years, although slightly over 4000 years would have elapsed back on Earth. If the ship were to return to Earth via the same route, upon their arrival back home the spacefarers would find that they were only 32 years older, but that 8000 years had passed by on Earth. All of the people that the space travelers had left behind would be long dead, and the original departure of the ship would be as remote an event to those witnessing its return as the origin of civilization itself is to us.

If a starship could accelerate at this same steady rate for an even longer time, one-way trips to any point in the galaxy could be made in less than 20 years of ship time. Fifty years of ship time would be sufficient to reach any point in the visible universe! Unfortunately, enormous energies are required to propel objects as large as starships to near-luminous velocities. The energy required to send a ship only 50 times the size of the Skylab space station on a 14-year (ship time) journey to a star 1400 light-years away is no less than 1×10^{27} joules. This is 10 million times greater than the total annual energy output

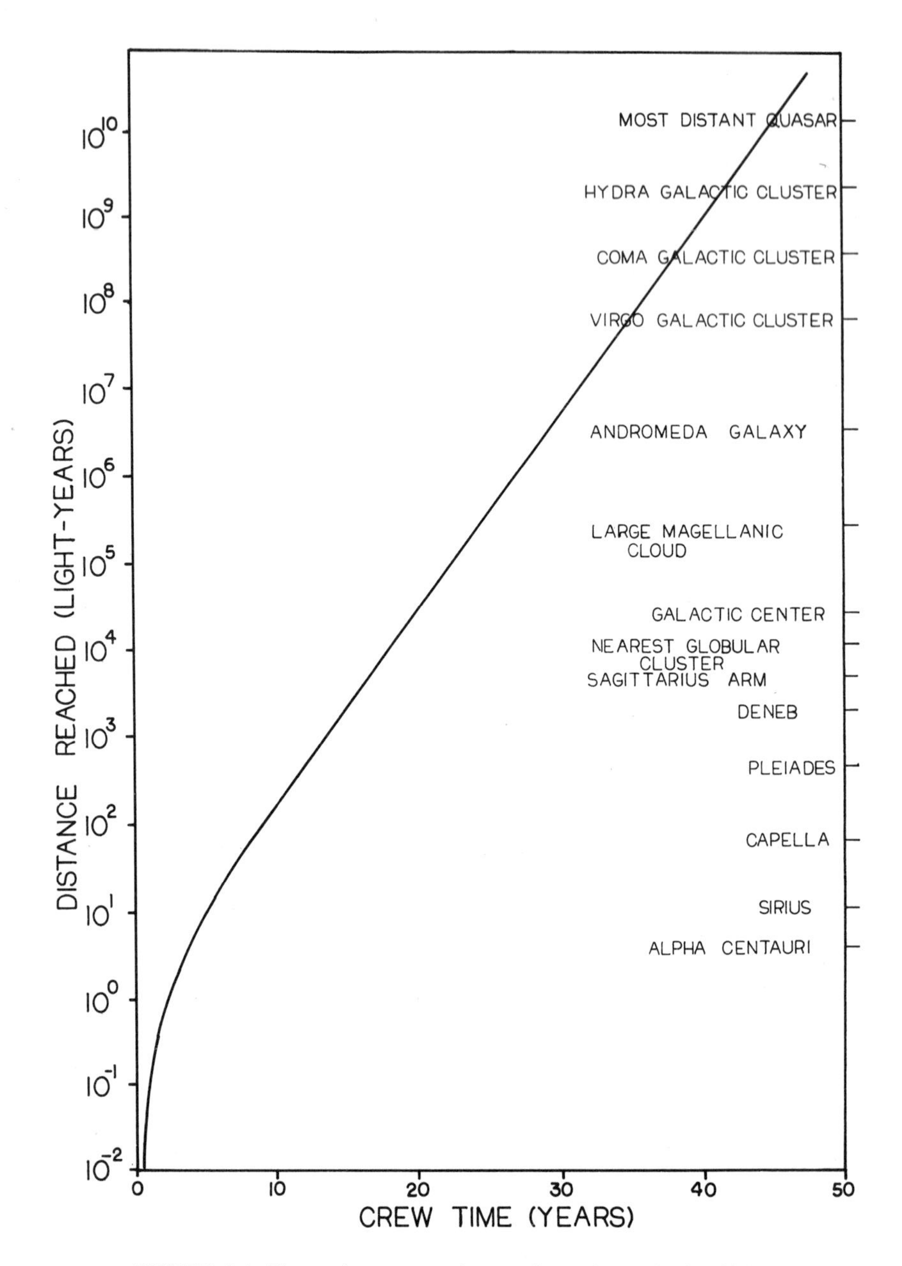

FIGURE 4.4. Time taken to travel to various places in the Universe, as measured by the crew aboard the rocket. It is assumed that the ship accelerates at a constant rate of 10 meters per second for the first half of the trip, then decelerates at the same rate for the remainder. Reprinted with permission from Sagan (1963). Copyright 1963, Pergamon Press, Ltd.

of all the nations on Earth. Societies capable of interstellar travel at relativistic velocities must of necessity be at least Type II civilizations.

Where could such enormous energy resources be obtained? Einstein's theory of relativity provides one answer. According to the theory, energy and matter are related by the famous equation

$$E = mc^2 \tag{4–2}$$

A truly enormous amount of energy is stored in matter: There is enough energy potentially available in a 1000-kilogram mass to run the entire world's economy for a full year. The problem lies in being able to find a way to extract this energy. In nuclear fission and fusion reactions some mass is converted into energy, but the conversion efficiency is less than a percent. If a way could be found to release *all* the energy stored in matter, a truly astronomical energy source would be available. The only way currently known to accomplish this is via a direct reaction between matter and antimatter. Every particle in the universe has a "mirror-image" antiparticle. Whenever matter and antimatter meet, they annihilate each other and release all of their stored energy in a flood of gamma rays which travel at the speed of light. A gamma ray–propelled starship whose mass is 90 percent matter-antimatter fuel could reach a velocity of 98 percent of light speed.

Unfortunately, antimatter does not occur naturally, at least not in our particular part of the universe. It must be created artificially in the laboratory, an atom at a time. Just as much energy must be expended to create the antimatter in the first place as is released when it is reacted. In order to be of any use as starship fuel, it would have to be synthesized by the thousands of tons. The manufacture of antimatter fuel on such a large scale would require access to virtually unlimited energy resources and would demand the creation of an entirely new technology. The storage, containment, and handling of

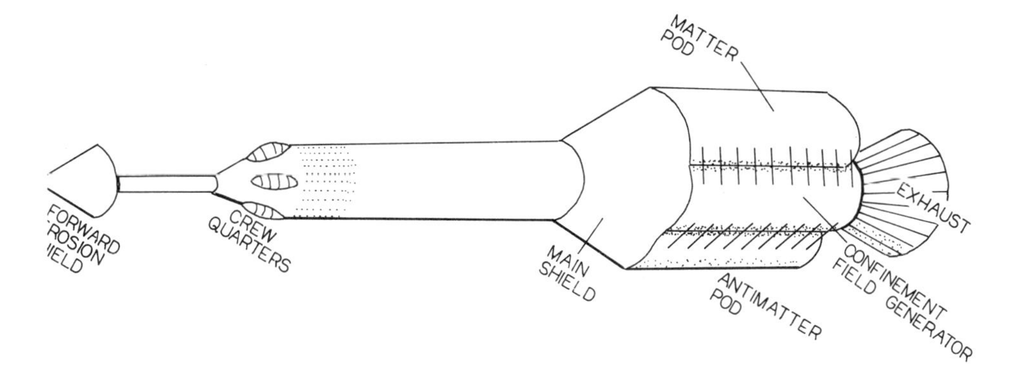

FIGURE 4.5. Starship powered by matter-antimatter fuel.

the antimatter while aboard the rocket are particularly important concerns. Antimatter must never be permitted to come into physical contact with ordinary matter before being directed into the reaction chamber; the slightest leak would trigger a catastrophic explosion which would probably be visible at interstellar distances. There is no known means by which gamma rays can be directed to form an exhaust. The mass of the shielding required to protect the crew from stray gamma rays would be immense. Even the high velocity of the ship would present totally different problems. At relativistic speeds, even a tiny grain of dust has as much energy as a rifle bullet. The diffuse interstellar hydrogen would produce a potentially lethal shower of radiation and secondary particles when the ship plowed through it at such high velocities. The problems involved in relativistic spaceflight seem so immense that the first practical matter-antimatter starship must lie many hundreds of years in the future.

LASER-PROPELLED STARSHIPS

One of the chief disadvantages of any type of rocket vehicle is that it must carry a massive amount of fuel as it travels into space. Is there a way to leave the fuel at home? This would be impossible unless the vehicle could be "pushed" from the ground. An ingenious way to do just that has been suggested. A craft would be built with a large, reflective "sail." A powerful ground-based laser beam would be directed upon the sail, exerting an optical pressure which would accelerate the ship away from the solar system. The concept first appeared in an article by Robert Forward, of the Hughes Aircraft Corporation, shortly after the laser itself was invented in 1960.

The technological challenges provided by such a laser propulsion sytem are immense. The laser must be enormously powerful and would have to be directed onto the sail for many years and over distances of many light-years in order for the ship to reach an appreciable fraction of the speed of light. The laser beam cannot diverge by any appreciable amount over the course of the mission. This requires that the transmitted beam be at least several hundred kilometers in diameter. The most powerful laser beams of today are scarcely larger than a few centimeters across. A high degree of precision will be required to keep the laser beam steadily directed onto the craft as it attains interstellar distances; a pointing error of less than a millisecond of arc could cause the beam to miss its target. Even with laser beams having hundreds of terawatts of power, flight times of several hundred years to the nearest stars can be anticipated. The people operating the laser system back on Earth will have to be exceedingly patient.

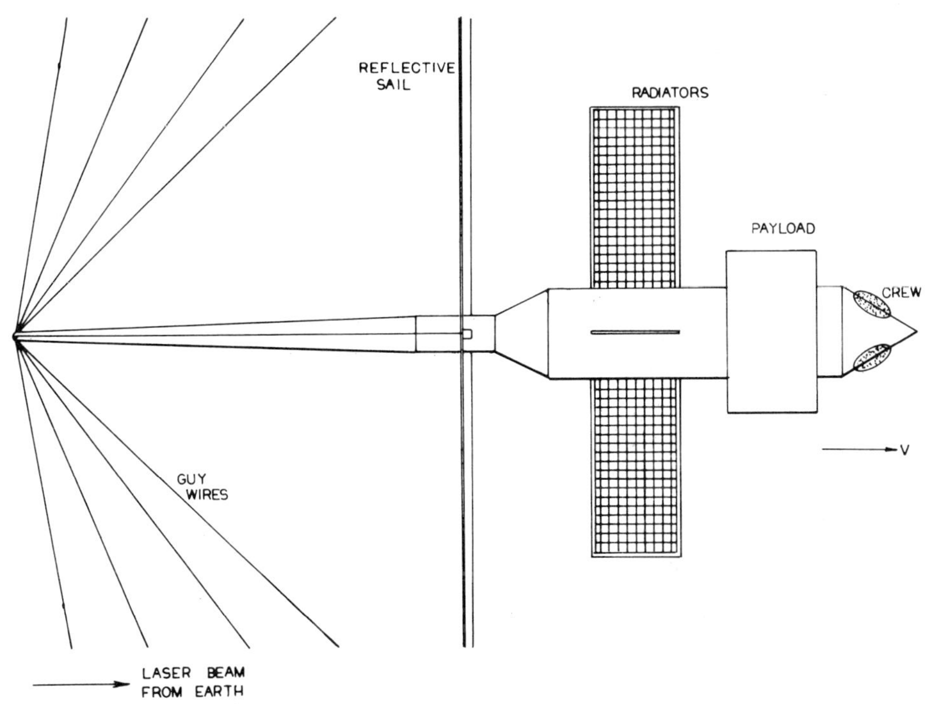

FIGURE 4.6. Laser-propelled starship. The craft contains a large "sail" which is pushed by a powerful laser on the ground.

Perhaps the most difficult problem of all to solve is that of the large sail which must be carried by the craft. In order to provide a sufficiently large accelerating force, the sail must be truly huge. It must be at least several thousands of square kilometers in area, nearly as large in diameter as the laser beam itself. The great advantage of the laser propulsion system is that the massive engines and fuel tanks required by conventional rockets need not be accelerated to relativistic velocities. However, it seems that we have eliminated the engines and the fuel only at the expense of requiring a massive sail. Most of the mass of the ship will have to be taken up by the sail, leaving precious little left over for payload.

INTERSTELLAR RAMJET

The prospect of reaching the stars within the natural lifetimes of the crew and passengers of a starship is rather remote if only nuclear or thermonuclear fuels are available. Less than a percent of the mass of these fuels is actually converted into energy; even the most advanced fusion propulsion schemes that have been proposed could probably reach velocities no greater than 20 percent of the speed of light. Laser-propelled vehicles would probably take thousands of years to reach even the nearest stars. Matter-antimatter propulsion appears to offer a real possibility of being able to reach a reasonable number of nearby stars within a few decades rather than millenia, but such technology requires so many radical advances as to appear all but impossible at present. All of these techniques suffer from the basic disadvantage that they uniformly require the expenditure of vast amounts of energy and the commitment of enormous resources. The effort required for successful high-speed interstellar space flight using any one of these propulsion schemes is so vast that it is probably feasible only for Type II or even Type III civilizations.

There may be another way to achieve relativistic space flight, one which does not require the expenditure of such enormous amounts of energy. In 1960 R. W. Bussard, an engineer working at the Los Alamos Scientific Laboratory, introduced a truly ingenious proposal: a high-speed craft which can use the tenuous interstellar hydrogen that it encounters on the way to the stars as its fuel. Such a vehicle is known as an *interstellar ramjet.* It would have a large "scoop" mounted on the front which would sweep up interstellar hydrogen as it passed through space and funnel it into a fusion reaction chamber. The trapped hydrogen would then undergo fusion reaction within the chamber, and the reaction products would be expelled from the rear of the craft to provide thrust. Since the average density of hydrogen in interstellar space is only a million atoms per cubic meter, the scoop must be hundreds of kilometers in diameter in order to gather up enough fuel to provide a thrust sufficient to reach relativistic velocities. Such an astronomically large scoop certainly could not be made from solid matter, but would rather consist of a complex pattern of electric and/or magnetic fields arranged to direct the interstellar hydrogen into the proper receptacle as the ship flew through space.

In order for an interstellar ramjet to function, the vehicle must first be accelerated to a relatively high initial velocity by conventional means. In a typical mission, Daedalus-type fusion engines would accelerate the ship to a velocity of 0.15c. The empty fuel tanks would then be dropped off, and the scoop would be deployed. The acceleration would be quite gradual at first, but it would steadily increase as the ship moved more rapidly through the interstellar medium. The ship could in principle continue to accelerate indefinitely, getting progressively closer and closer to the speed of light. The craft could reach virtually any point in the galaxy within 150 years of ship time, although several hundred thousand years would pass by back on Earth.

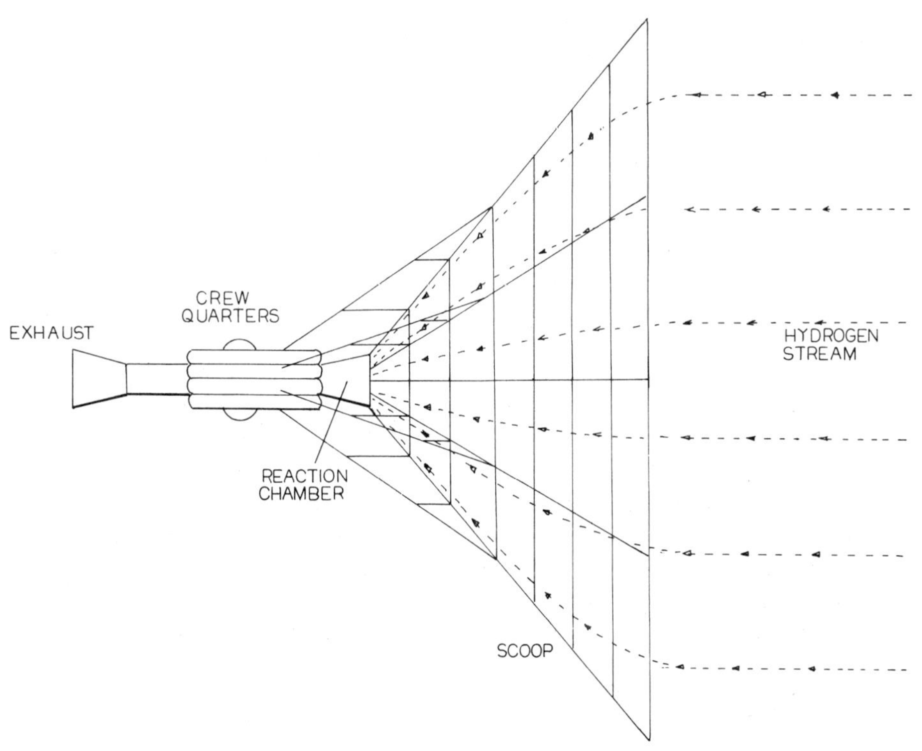

FIGURE 4.7. Bussard interstellar ramjet. From R. L. Forward (1976). Copyright 1976, British Interplanetary Society.

The particular beauty of the interstellar ramjet is that its construction and launching would require far less initial investment of energy and material resources than other advanced, high-speed space propulsion systems. There are, of course, many technological problems which must be overcome before such a ramjet is feasible, but the idea of achieving relativistic space flight by picking up the fuel along the way is such an intriguing idea that it deserves much more study. It is worth quoting from Bussard's original paper at this point:

> This (hydrogen scoop) is very large by ordinary standards, but then, on any account, interstellar travel is inherently a rather grand undertaking, certainly many magnitudes broader in scope and likewise more difficult than interplanetary travel in the solar system, for example. The engineering effort required for the achievement of successful short-time interstellar flight will likely be as much greater than that involved in interplanetary flight as the latter is more difficult than travel on the surface of the Earth. However, the expansion of man's horizons will be proportionately greater, and nothing worthwhile is ever achieved easily.

TABLE 4.2 Summary of Various Modes of Interstellar Transport

Type of Rocket	Exhaust Velocity (m/sec)	Maximum Feasible Velocity (m/sec)	Travel Times: Inner Planets	Outer Planets	Nearest Stars	Stars 1000 l-y distant	Edge of Galaxy
Chemical	4×10^3	15×10^3	Several months	Several years	Thousands of years	Millions of years	Inaccessible
Nuclear Fission	6×10^6 $= 0.02c$	2×10^7 $= 0.06c$	A few days	A couple of weeks	Several hundred years	Several thousand years	A million years
Nuclear Fusion	3×10^7 $= 0.1c$	4.5×10^7 $= 0.15c$	A few days	One week	A few decades	Ten thousand years	Several hundred thousand years
Laser	none	0.98c	Not feasible	Not feasible	Several centuries	A thousand years(*)	Ten to twenty thousand years(*)
Matter-Antimatter	c	0.999c	Not feasible	Not feasible	A few years(*)	Ten to twenty years(*)	A thousand years(*)
Interstellar Ramjet	c	arbitrarily close to c	Not feasible	Not feasible	Ten to twenty years(*)	Thirty to forty years(*)	150-300 years(*)

(*) Proper time, with respect to the crew aboard the rocket.

MESSENGERS FROM THE STARS

In 1960 Ronald Bracewell, a radio astronomer working at Stanford University in California, proposed that the most efficient way for a race of intelligent beings to search for others like themselves would be to send automatic "messenger probes" to the vicinity of those nearby stars that show potential of having life-sustaining planets. Upon arrival, a probe would enter orbit within the ecosphere of the star and carry out a detailed search for a technological society. If the probe finds such a civilization, it could initiate contact. Bracewell proposed that an alien messenger probe might be present in our solar system at this very minute, quietly studying the human race as it emerges into the communicative phase.

Messenger probes must be capable of operating for thousands or millions of years very far from home. Such a probe would require sophisticated systems capable of automatically repairing any damage or malfunction. Since the two-way communication time between a probe on station and its creators back home is likely to be many decades or even centuries, it is impractical to direct its day-to-day operations via radio telemetry in the same way as terrestrial space probes flying to nearby planets are controlled from Earth. The probe must be able to act upon its own volition, without having to wait for orders from home. This would require that the probe have a high degree of intelligence, at least the equal of the human brain and in all probability far superior.

Gerard K. O'Neill, of Princeton University, has proposed that messenger probes launched by a highly advanced civilization might even be self-replicating, much like an amoeba. When such a probe arrives in a new solar system, it enters orbit around the star and immediately begins to direct the construction of exact duplicates of itself out of materials available locally. Once the copies are completed, they are sent on their way to other star systems while the original remains behind to search for intelligence. Probes capable of self-replication in this manner could be considered as being "alive" and, indeed, highly intelligent beings in their own right. The famous mathematician John von Neumann demonstrated that such self-replicating machines are theoretically possible. Automatic self-replicating probes could continue to advance throughout the stars long after the civilization that originated them has vanished.

A messenger probe could search for intelligent activity via several different means. The probe could listen for narrow-band radio signals that might be produced by a technological society similar to ours. The probe could also be equipped with optical devices to examine the surfaces of likely planets for signs of large-scale engineering activity. This would ensure that civilizations which fail to evolve radio technology at an early stage are not overlooked. As additional insurance, the probe might even be equipped with automatic landing craft to explore planetary surfaces from close up.

Once a messenger probe is satisfied that a communicative technology is present, what does it do next? A premature revelation of the probe could be catastrophic for both the society and the probe alike. Some new societies may not be intellectually or psychologically prepared to accept the existence of intelligent aliens. A worldwide wave of panic and hysteria could be the first result of the probe's discovery. An alien spacecraft's sudden appearance in their immediate neighborhood might be misinterpreted by the new society as an impending invasion, and the probe might even be attacked and destroyed. The famous *War of the Worlds* radio broadcast made by Orson Welles in 1938, which induced large-scale panic over much of the east coast of the United States, is an example of what could happen worldwide if the probe is prematurely discovered. The new knowledge that such a sophisticated probe would suddenly make available might be highly dangerous to an immature society. The probe's technology might fall into the hands of one particular nation or group, ultimately to be used to exploit or conquer others. The probe might even encounter an advanced civilization which is viciously militaristic. The launching society itself could be endangered. Because of these risks, messenger probes will almost certainly be designed to operate so that they cannot be inadvertently detected. The messenger probe will make careful, clandestine studies of any civilization that it finds before it reveals its existence or attempts contact.

The probe may be instructed by its builders to avoid any interaction with civilizations that are divided into squabbling nation-states, torn by religious strife, faced with economic and social conflict, or otherwise disunified. The probe may be ordered not to attempt contact even with a totally unified global civilization if it finds such a society to be militaristic, repressive, xenophobic, superstitious, or otherwise repugnant. These societies may be a danger to others as well as to themselves. Human civilization may at this very moment be undergoing interstellar judgment on the basis of its radio and television signals. The first impression that extraterrestrials have of the human race may be from soap operas, situation comedies, fanatical radio preachers, rock-and-roll music, and the never-ending pattern of violence and sensationalism of television news.

Once the probe does decide to attempt contact with a civilization, several different strategies are open. The probe could simply transmit a coded "beacon" signal at regular intervals and patiently await a response. Bracewell proposed that one means of contact would be for the probe to record any stray intelligent radio signals that it happened to pick up and beam them right back to their source. This would continue until the probe is satisfied that its presence has finally been noticed. It could then transmit information about its origin and intentions to the target civilization. The first message might be a picture of the sky as seen from that particular star system, showing the probe's origin. Further information could be sent once a two-way dialogue has begun. At this time, the probe might send the word back home that an alien intelligence has at last been spotted.

Bracewell suggests that a serious search be made for a messenger probe in our own solar system. Strangely enough, there have been sporadic reports over the years of odd radio echoes having delays as long as a few minutes. The most widely known of these reports was by Stormer and van der Pol in the late 1920s. They noted that short bursts of radio pulses beamed from radio station PCJJ in Eindhoven, Holland were being faithfully repeated 3 to 15 seconds after their original transmission. A similar sequence of "delayed echoes" was noted in France in 1929. These echoes have never been adequately explained. Could they have come from an alien messenger probe? In 1972, a Scottish investigator by the name of Duncan Lunan proposed that the Stormer–van der Pol delayed echoes were actually coded instructions for the drawing of the constellation Bootis trans-

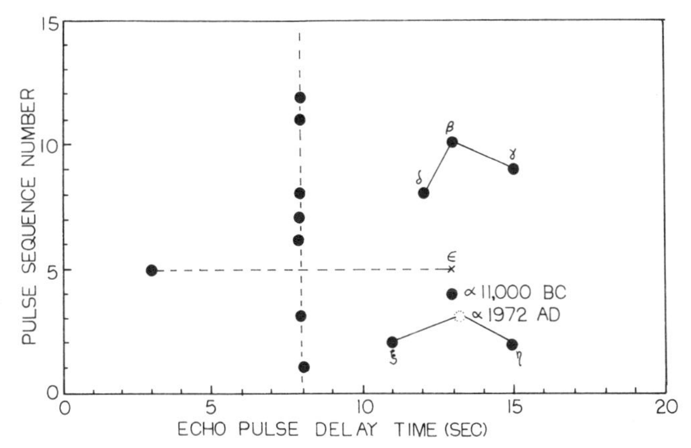

FIGURE 4.8. Duncan Lunan's interpretation of the Stormer/van der Pol sequence of delayed echoes. From A. T. Lawton (1974). Copyright 1974, British Interplanetary Society.

mitted by a probe located near the Moon. The bright star Arcturus is in this constellation, but the probe came from the epsilon member, since it is displaced in the diagram. Judging from the small differences between the positions of the stars in the diagram and those in the actual constellation, Lunan estimated that the probe arrived about 13,000 years ago. The French sequence of echoes gave even more information. The civilization that built the probe originated on the second of a seven-planet system, but they had found it necessary to move outward to the sixth planet because their home star was nearing the end of its main-sequence life. Lunan's analysis was greeted with considerable skepticism by the scientific community. Delayed echoes are indeed noted from time to time, but never the same sequence of delays twice. They appear to be random events, and other experimenters have not been able to reproduce the same sequence of pulses noted by Stormer and van der Pol.

Another incident that is often discussed in books and articles on interstellar travel and extraterrestrial intelligence is the reported reception in England of television signals sent from the United States three years earlier. This is the famous "KLEE Incident," which is often reported as fact and has entered the folklore of the mysterious along with the Bermuda Triangle. According to the story, on one evening in September of 1953 British television viewers were surprised to see the test pattern of television station KLEE of Houston, Texas appearing on their screens. This event in itself was rather unusual, since television stations only rarely can be received at ranges greater than a couple of hundred miles. Several viewers were curious enough to photograph the screen and mail the pictures to the Texas-based television station. It was even more startling when it was found that KLEE had gone off the air three years earlier! It had been sold to another group in 1950 and had changed its name to KPRC. The story spread, and the incident was investigated by researchers on both sides of the Atlantic. To this day it is reported as a great mystery. The truth of what actually happened has, however, been known for a long time. Frank Drake tells the story. It seems that a group of confidence tricksters in England had been promoting a new type of television receiver claimed to be capable of picking up direct signals from American television stations. Since this was impossible at the time (and still is), the group had to resort to fraud. First, they came to the United States and took photographs of television screens displaying test patterns. Then they projected these images onto the screen of their device, giving British viewers the impression that the receiver was actually picking up an American television station. Potential investors were encouraged to write to the American television stations for confirmation. One of these stations happened to be KLEE, and the tricksters had apparently forgotten to check to see if it was still on the air. Thus the beginning of a legend.

Recently, Hiromitsu Yokoo, of the Kyorin University in Tokyo, and Tairo Oshima, of the Mitsubishi-Kasei Institute of Life Sciences in Tokyo, proposed that some Earth organisms might actually contain a coded message from an extraterrestrial intelligence, stored perhaps in the base-pair sequence within their DNA molecules. A likely candidate for the biological messenger might be the phage

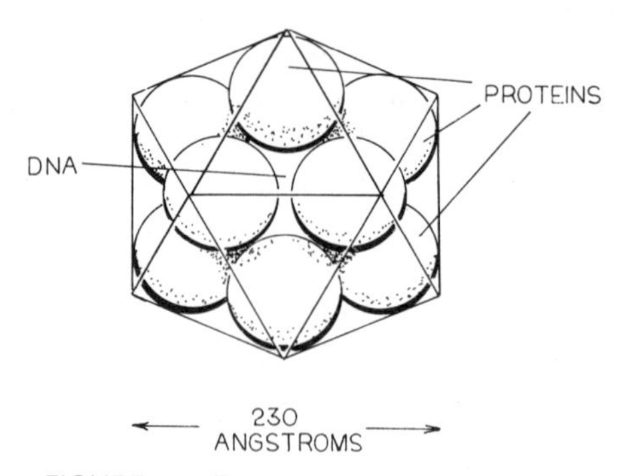

FIGURE 4.9. The bacteriophage ϕX-174.

ϕX-174, one of the viruses which infects E-coli bacteria in humans. Viruses occupy the "gray" region between the living and the nonliving. The simplest virus consists of little more than a large, double-stranded DNA molecule surrounded by a protein coat. It is capable of reproducing itself, but only by usurping the genetic apparatus of the organism it infects. Left without a host, a virus appears to be no more alive than a stone. A virus can in a certain sense be regarded as the very simplest and most primitive type of microorganism. Perhaps many millions of years ago the sequence of base-pairs in the DNA of a virus was intentionally altered by alien visitors to Earth so that it would spell out some sort of message, perhaps a coded instruction for the drawing of a two-dimensional picture giving a description of their civilization. The new organism was then reintroduced into the terrestrial ecology and allowed to multiply. The message would then be passed on to succeeding generations until intelligent beings at last evolved the capability of noticing and deciphering it.

An entirely different strategy would be for a messenger probe to construct a vault filled with information and artifacts in the target system and then press onward to other stars. This would avoid the need for the probe to wait the millions of years it could take for a civilization to emerge. The vault could contain a description of the civilization that sent the probe, as well as a set of instructions for the initiation of contact. There might even be a sort of cosmic "burglar alarm" that would alert the builders that a civilization had appeared capable of opening the vault and examining the contents. The vault could be made of materials available locally, but it would have to stand up against the ravages of millions of years of time. It may be unwise to place the vault on an earthlike planet where intelligent life might emerge. Erosion or geological activity would probably destroy the structure long before it is ever found. In addition, the vault could be discovered by intelligent beings before they have attained sufficient maturity to appreciate its contents. Driven by ignorance, fear, or superstition, they might destroy the vault and its precious contents would be irretrievably lost. A better location might be an airless, geologically dead moon. Such worlds have surfaces that remain essentially unaltered for billions of years. This strategy would make the vault accessible only to an intelligent race that had developed interplanetary space travel and showed some capability for long-term survival. There could be such a vault in our own solar system, possibly located on one of the moons of Mars, on one of the moons of Jupiter, or on some of the frozen moons of the outer planets such as Saturn, Uranus, or Neptune. Future space missions should keep this possibility in mind.

A MESSENGER FROM EARTH

The first man-made object to be launched toward the stars was the space probe Pioneer 10, launched from Cape Canaveral on March 2, 1972. It passed within 2 planetary radii of Jupiter, and during the encounter it picked up enough additional velocity to escape from the solar system with a residual velocity at infinity of 11.5 kilometers per second. The probe will cross the orbit of Pluto in 1989 and is heading approximately toward the star Aldebaran in the constellation Taurus, which it will reach approximately 1.6 million years from now.

Although there is a negligible probability that Pioneer 10 will ever enter another solar system, it was nonetheless thought appropriate that the first man-made object to enter interstellar space should carry some sort of symbolic message to extraterrestrial civilizations. Shortly before the scheduled launch, Carl Sagan and Frank Drake, of Cornell University, suggested that a small plaque be placed aboard the Pioneer spacecraft that would give a description of the location and general appearance of the human race. Since the weight of the proposed plaque was small, NASA readily agreed. The most striking feature on the plaque is, of course, the representation of the human male and female. They are shown standing in front of a schematic drawing of the Pioneer 10 spacecraft, so that aliens can get an idea of their relative sizes. An attempt was made to avoid racial stereotypes, so that the man and woman could be considered as representative of all humankind. The man has his right hand upraised in what is intended as a human gesture of greeting, showing at the same time that his hand has four fingers and an opposable thumb. It was decided that it was not a good idea to show the couple holding hands, lest the extraterrestrials infer one creature rather than two.

The two circles on the upper left of the plaque are meant to indicate a distance and time scale. They are intended as a schematic representation of the hyperfine transition in atomic hydrogen which is responsible for the microwave emission that takes place at a wavelength of 21 centimeters or 1420 MHz. This radio noise should be well known to all technological civilizations which have at least reached the level of astronomical sophistication attained by current-day Earth inhabitants. Just below the diagram is the binary digit 1, indicating that the hydrogen hyperfine transition somehow represents a unit of either space or time. The most logical choice for the unit of length would be the wavelength of the hydrogen emission (21 centimeters); the most logical unit for time the period of a single hydrogen radio oscillation (70 nanoseconds). In order to help the extraterrestrials confirm that this is indeed the correct interpretation, the height of the female is indicated by the binary equivalent of the

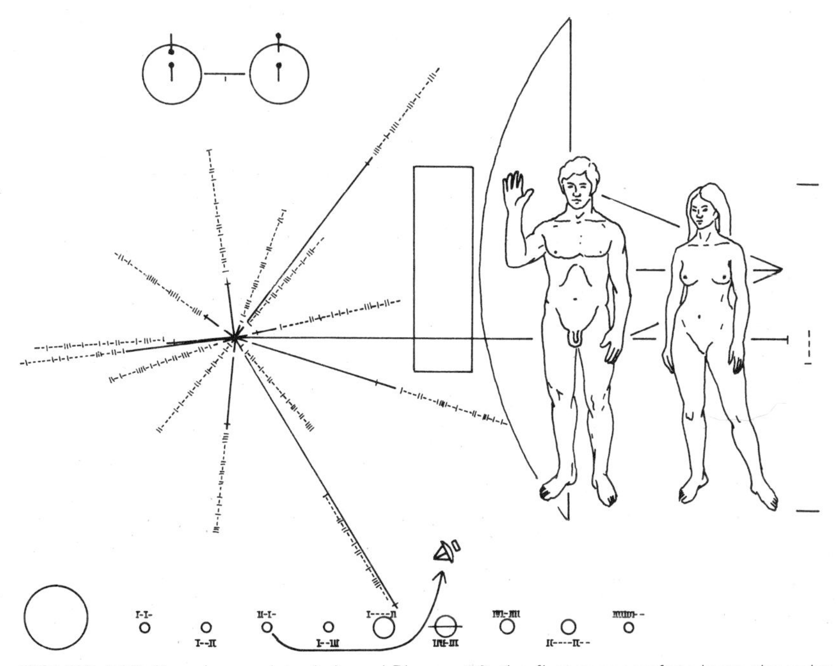

FIGURE 4.10. The plaque placed aboard Pioneer 10, the first spacecraft to leave the solar system. From Sagan, Sagan, and Drake (1972). Copyright 1972 by the American Association for the Advancement of Science.

number 8, showing that her height is 8×21 cm = 168 centimeters or 5 feet, 6 inches.

At the center of the plaque is what appears to be some sort of polar coordinate representation of the positions of some objects relative to an origin. This figure is designed to tell the extraterrestrials where we are. There are fifteen lines emanating from a single point, with fourteen of these lines having a large binary number (of the order of a hundred million) attached. These numbers could conceivably be either space or time units. If they are space units, the numbers designate distances of the order of 30,000 kilometers. If they are temporal units, they represent time intervals of the order of 0.10 second. Which is the correct interpretation? Although there is nothing in the figure that would otherwise assist an extraterrestrial in making a choice, the authors of the figure intended the numbers to represent the periods of known pulsars. The figure thus shows the positions of several pulsars relative to an origin. This origin could conceivably be the galactic center or it could be the Earth itself. A clue is given by the inclusion of a fifteenth line without any number attached to it. This line is intended to represent the distance from the Earth to the center of the galaxy and provides a distance scale for the other lines in the diagram. The lengths of the lines therefore represent the distances between the various pulsars and the Earth, measured in units of the Earth-to–galactic center distance. Note that there are tic marks located near the ends of the fourteen pulsar lines. The distance between the end of the line and the tic mark is meant to represent the distance above or below the galactic plane that the pulsar is located.

At the bottom of the plaque is what is obviously a schematic representation of the solar system, with the Sun at the left. The spacecraft is shown as having originated from the third planet and as having flown past the fifth. Near each of the planets is shown a binary number to indicate its relative distance from the Sun. These numbers have serifs on them to indicate that the distance scale used here is not the same one that was used to specify the sizes of the humans. This new distance unit is one-tenth of the semimajor axis of Mercury, as the binary representation of ten (1010) is shown next to that planet. There is no way that this fact could be discerned from the message itself, but it is rather unlikely that intelligent extraterrestrials would confuse this distance scale with that derived from the hydrogen radio emission, as that would require a solar system with Mercury only 210 centimeters from the Sun!

Any civilization that is capable of discovering and intercepting the spacecraft in the first place should be clever enough to figure out the rather complicated code and decipher the message. They presumably would have complete records of the position, proper motion, and period of virtually every pulsar in the galaxy. Their star charts should include precise information about the type, location, and proper motion of every single star in the galaxy within a range of 1000 light-years of their home planet. Armed with such data, a detailed analysis of the plaque should permit them to pinpoint the location of our Sun to within 60 light-years or so. The diagram of our solar system at the bottom of the plaque should give enough information so that it could be uniquely identified among the star systems in its immediate vicinity. Since pulsars "run down" at known rates, the extraterrestrials could also determine the approximate time that the Pioneer spacecraft was launched from Earth. Once they know where our planet is located, they could then attempt to establish contact, either by sending a radio message or by launching an expedition to Earth. There were several scientists who expressed some apprehension about revealing Earth's location to extraterrestrials in this manner, since we could be exposing ourselves to military conquest or economic exploitation at their hands. The danger of Earth falling victim to interstellar imperialism is minimal; there is probably not much that we humans have that aliens could not obtain far more easily much closer to home. We can only hope that a technological civilization capable of acquiring the Pioneer spacecraft in the first place will also be of sufficiently high moral character to regard the plunder of our planet as unthinkable.

It will be at least several million years before any civilization will be able to intercept the Pioneer spacecraft. By that time, the human race certainly will have changed beyond all recognition, and any sociological or biological information inferred from the plaque will be obsolete. We human beings may have disappeared from the Earth altogether, having destroyed ourselves in fratricidal conflict many years earlier. Even if our remote descendants are still here, they will undoubtedly have evolved into quite different creatures from twentieth-century man. Perhaps in the far distant future we humans will have been replaced by superintelligent machines based on metals and semiconductors rather than carbon. The race of beings that finally intercepts the craft could conceivably be descendents of humans who originally left Earth to colonize the far reaches of the galaxy many millions of years earlier. All memories of the motherland may have been completely lost or else reduced to vague and fantastic legends of an incredibly ancient past. It is interesting to speculate about the impact the discovery of the Pioneer spacecraft will have on the society that finds it.

Carl Sagan collaborated with Frank Drake, A. G. W. Cameron, Philip Morrison, Bernard Oliver, and Leslie Orgel to place a different sort of message aboard the later Voyager spacecraft that were scheduled to fly past Jupiter and Saturn on their way out of the solar system. This time the message was carried by a long-playing record. The record was made of copper and was placed in a protective sleeve. A cartridge and needle were provided, along with a set of instructions on how the record is to be played. The record consisted of several electronically encoded pictures that show such scenes as the Earth from space, terrestrial landscapes, natural wonders, various human activities, engineering achievements, aircraft, space vehicles, scientific instruments, and radio telescopes. There were samples of 55 spoken languages included, as well as an encoded message from President Jimmy Carter. There was a selection of Earth sounds, ranging all the way from gentle wind and surf to the deafening roar of the launch of a Saturn 5 moon rocket. Finally, a selection of music was provided, chosen from many different Earth cultures. These cover all sorts of music, ranging from Navajo chants to the classical works of Bach and Beethoven, from African drums to the jazz of Louis Armstrong and the rock-and-roll of Chuck Berry. The record is an attempt to portray the whole breadth of human experience to extraterrestrials. What will they make of it?

INTERSTELLAR COLONIZATION

There is apparently no law of physics which specifically forbids interstellar travel, even at speeds very close to that of light. Within the foreseeable future it may even be possible to send forth manned missions to nearby star systems that will reach their destinations within times as short as a few centuries. Enormous "generation ships" may ultimately venture forth from Earth carrying sizeable numbers of human beings to carve out new homes on planets around other stars. Assuming that humanity survives the crises that currently threaten its continued existence upon Earth, it could eventually spread its seed throughout the galaxy until every available niche is occupied. The human race itself may be destined to evolve into a galactic Type III civilization a few million years hence.

Eric Jones, of the Los Alamos Scientific Laboratory, has simulated the future expansion of the human race into the galaxy by using a "Monte Carlo" computer model. He assumes that interstellar colonization will take place according to a pattern similar to the Polynesian settlement of the remote islands of the Pacific. Included in the model are such limiting factors as control on human populations and the finite velocities attainable by starships. The mean separation between habitable planets is taken to be 7 light-years, and the starships which carry the colonists to their new worlds are assumed to travel at 0.10 light speed. Once a new colony is established, its population will grow to half the saturation value within 2200 years. At this time the high population density will force the inhabitants to send out another wave of starships in search of new worlds to inhabit. Under these assumptions, the computer calculation predicts a spherical wave of human colonization spreading outward from Earth into the galaxy at an average speed of approximately 3 light-years every 2000 years. At that rate, the galaxy could be completely filled with the evolutionary descendants of humankind within 56 million years, an incredibly short time on the cosmic scale. Within less time than it took for the first prosimian mammals to evolve into humankind and conquer the Earth, the descendents of humans will spread out to completely dominate the entire galaxy. Assuming each habitable planet to be capable of supporting 20 billion inhabitants, the galaxy might eventually be home for 7×10^{19} human beings less than a hundred million years from now.

Frank Tipler, of Tulane University, has suggested that there is a particularly economical manner by which the human race could colonize the entire galaxy by launching just one single starship. This would be possible if that starship were a von-Neumann type of automated "universal constructor," capable of synthesizing anything its designers chose, even including an exact duplicate of itself. This superintelligent ship would be launched from Earth and directed toward a nearby star system with the goal of founding a human colony there. Upon arrival at its destination, the ship would build a series of structures intended for human habitation on the surface of a suitable planet and out of materials available locally. The newly prepared colony could be populated by humans grown from pre-fertilized egg cells carried from Earth or freshly synthesized upon arrival. The newborn infants would be raised by the ship's computer until they were able to occupy their new home. During this time, the ship would direct the manufacture of exact duplicates of itself and its cargo. Upon completion, these copies would be launched toward other stars in search of new homes for humankind, gradually extending human habitation outward into the galaxy until all accessible niches are occupied. The particular beauty of this scheme for interstellar colonization is that the cost of the design, construction, and launching of the *first* universal constructor starship is the only expense incurred by Earth. In fact, the wave of human galactic colonization could proceed unimpeded even if the Earth itself were to perish.

The colonization of alien worlds is not likely to be an easy task. Every habitable planet in the galaxy will have some form of life. The ecologies of these worlds are certain to be highly toxic to humans. People can eat the food, drink the water, and breath the air here on Earth only because they are a product of nearly 4 billion years of continuous evolutionary development on this particular planet. Human beings may be unable to adapt to the biospheres of totally alien worlds. Instant death may await anyone foolish enough to walk out unprotected onto the surface. In order to survive on such a toxic world, human colonists would have to construct artificial environments which keep people isolated from the air, water, and soil of their new home. If they ever wished to venture outside their protective bubbles, the colonists would first have to totally destroy the entire ecology of their new world and replace all of the indigenous life forms with those brought along from Earth.

If past evolutionary history is any guide, our human descendants who colonize the habitable worlds of the galaxy a few million years hence will be as different from us mentally and physically as we are from our ape forebears. Each habitable planet has a different environment and will favor the evolution of a different type of human being. The slow accumulation of gradual evolutionary adaptations which takes place over many millenia may ultimately result in such major changes that the remote descendants of the original settlers will no longer even be genetically compatible with earthlings or with the inhabitants of other colonies, should physical contact ever be established. In this way, myriads of new human species will appear, each uniquely adapted to its own special environment.

Even though there could well be an intragalactic human culture millions of years hence, the interstellar empires of popular fiction will probably never appear. The mean separation between habitable worlds is expected to be approximately 25 light-years, which means that it will take many decades for an exchange of messages between nearest neighbors, and perhaps several centuries for even a one-way trip. Any sort of interstellar government may be impossible under such circumstances. A centralized ruling group may be unable to exert any effective authority over other worlds if there is a time lag of hundreds or even thousands of years between the transmission of orders and their receipt. Demands from the motherland are likely to be ignored if they are thousands of years out of date and especially if there is no possibility of any immediate reprisal. Long voyages between the stars may be feasible for initial colonization, but completely impractical for routine travel or trade. Once a colony becomes established on a new world, it will be completely independent of the rest of the galaxy and must be able to survive on its own. Should the new inhabitants manage to get themselves into trouble, any type of aid would be centuries away. There is, however, a real sense of security in being so remotely separated from one's neighbors; predatory activities such as space piracy, economic exploitation, enslavement, or military conquest over these extreme distances will probably not be worth the meager results that could be obtained. The inhabited worlds of the galaxy will forever be safe, behind the impenetrable barrier of the speed of light from their neighbors' lust for ill-gotten gain.

THE FERMI PARADOX

The human race might be able to establish a significant presence in the galaxy within a time span as short as a few million years, an incredibly short time on a cosmic scale. What we humans may shortly be able to do, advanced extraterrestrial civilizations should also be able to accomplish. Several people have reflected upon the fact that if large numbers of advanced technological civilizations have actually existed in the galaxy for as long as a few billion years, it appears at first sight rather odd that not a single one of them has ever attempted to colonize the Earth. If just one single ancient spacefaring civilization had begun to colonize the galaxy as recently as a few million years ago, all of the habitable worlds in the galaxy should now be occupied by descendants of their race. In virtually every case these alien colonists would have found it necessary to "terraform" their new homes to conform to their own pattern of life, in particular completely eliminating any sort of indigenous life which could prove to be dangerous. Every habitable planet in the galaxy should become completely and totally dominated by the descendants of the *first* intelligent race to undertake the colonization of the stars. Extraterrestrials rather than human beings should currently be living on the Earth. Since it is obvious that they are not here, where then are they? This problem has come to be known as the "Fermi Paradox," after the physicist Enrico Fermi, who was reported to have mentioned it while answering a question about the possible existence of intelligent extraterrestrials during an after-dinner speech that he gave during the 1940s.

One obvious resolution of the paradox is, of course, that extraterrestrial technologies do not exist. Many other worlds may have life, and a few may even have evolved highly intelligent creatures capable of abstract thought, but the Earth happens to be so different and unique a place that no other planet has ever produced a communicative society capable of interstellar space flight. The human race is alone, but if we survive we may someday dominate the entire galaxy. One of the most vocal exponents of this point of view has been Frank Tipler. He argues that once a technological society develops the capability for interstellar space flight, there will be intense pressures forcing it eventually to use this technology to expand outward to other worlds. These pressures might include such familiar physical problems as population overcrowding, diminishing energy resources, or high levels of pollution within the home solar system. The psychological pressures acting on completely alien intelligences are, of course, impossible to determine, but they could include such diverse drives familiar to us human beings as a desire for more living space, a striving for freedom from cultural, political, or religious restraints, an insatiable curiosity about other worlds and other places in the universe, a thirst for adventure, a need to seek out other intelligent beings, perhaps even a quasi-religious desire to extend the benefits of technology and civilization to every accessible niche in the galaxy. If for no other reason, every society will ultimately be forced to seek out new territory by such disasters as a nearby supernova or the imminent departure of its own star from the main sequence. Tipler believes that the current absence of aliens on Earth is so powerful an argument for their complete nonexistence that any sort of organized search for extraterrestrial technologies would be a waste of time.

There are alternative explanations which do not require the nonexistence of intelligent extraterrestrial technological societies. All of them are of necessity highly speculative. One possibility is that the human race is among the earliest technological societies to have appeared in the galaxy. Someone obviously had to be the first. Why not us? Several other societies could be ahead of us technologically, and the spherical wavefronts of their expanding civilizations may currently be advancing through space toward Earth, but none has yet reached our planet. However, they could arrive on Earth at any time. Alternatively, we may be destined to encounter the vanguard of another civilization as we ourselves advance outward into the galaxy.

Perhaps many millions of advanced technological civilizations have appeared in the galaxy, but the lifetime of every single one of them has been short. None lasted in the communicative phase any longer than a few centuries or at best a few millenia before they disappeared from the scene forever. For one reason or another they all managed to destroy themselves at the Type I level before they had a chance to develop any capability for large-scale interstellar travel. If intelligence has existed in the stars for several billion years, as many as a hundred million technological societies could have evolved. Every single one of them must have perished if the "self-destruction" hypothesis is to be a viable explanation for the current absence of extraterrestrials on Earth. All that would be necessary for the complete colonization of the galaxy is for just one civilization to have survived technological adolescence. If every previous society has indeed perished in such a manner, our own chances for survival must be very slim.

Maybe all of the advanced extraterrestrial societies that have come before us *voluntarily* chose not to expand outward into the galaxy. Perhaps all mature communicative societies tend over the years to lose the expansionist urges that are undoubtedly a part of their evolutionary heritage. It may be true that those few alien societies which survive to reach Type II status do so only by gradually or even purposefully eliminating from their psyches those ancient urges which would otherwise drive them to self-destruction. The desire to expand may have disappeared; they may be perfectly content to remain forever within their own solar system, eternally sealed inside their Dyson spheres and concentrating their efforts on activities, such as philosophical contemplation or personal development, which do not require access to more space or the consumption of larger amounts of energy. For the "loss-of-interest" hypothesis to be an explanation for the Fermi Paradox, it would have to apply to every civilization which has come before us. Many societies may lose interest in expansion, but it is unrealistic to expect that *all* would do so. All that is required for the colonization of the galaxy is for just *one* civilization to have retained its outward expansive needs and desires after it had reached the starfaring stage.

Perhaps extraterrestrials have not taken over the Earth because our planet has nothing that they can use. The conquest of another world is an extremely difficult undertaking and is likely to require many millions of years, even for the most advanced alien technology. Since the terrestrial biosphere is likely to be highly toxic to their form of life, the Earth may be no more tractable for extraterrestrial inhabitation than the lifeless, hostile planets like Venus or Mars or even the moons of Jupiter and Saturn. As a result, alien interstellar voyagers may have found it more practical to construct artificial O'Neill–type colonies in space. Such colonies could be assembled within a few years, whereas it could take millions of years to conquer a planet such as the Earth. Asteroid belts might be especially attractive places for such colonies because of the ready supply of raw materials. Michael Papagiannis, of Boston University, has suggested that perhaps there are aliens currently living in our own asteroid belt, housed within giant, hollowed-out planetoids. They have founded a highly stable society which has lasted for many millions of years. Quite unexpectedly, they now find that another technological species has appeared in their immediate neighborhood. They are keeping out of sight, being quite at a loss about what to do with us.

Sufficiently advanced races of intelligent creatures may inevitably even outgrow their requirement for a star about which to make their homes. Evolutionary pressures within populations which choose to undertake the long-duration voyages necessary for travel to other stars may tend to favor the development of creatures which are particularly well adapted for living in interstellar space, so well adapted in fact that they choose to remain there forever rather than to settle down near a star. They may be quite content to live a perpetual nomadic existence in the galaxy, eternally drifting aimlessly from one star system to another, stopping only long enough to replenish their fuel supplies and to pick up more raw materials. The scars left over from the mining expeditions of advanced extraterrestrials who passed through our solar system many eons ago may await us in the asteroid belt or on the moons of Jupiter or Saturn.

The "zoo hypothesis" is another suggested explanation for the paradox. Alien technological societies have failed to colonize the Earth because of a galactic moral imperative which prohibits them from interfering with any world where indigenous life is present. The act of wholesale ecological destruction needed to adapt the Earth to their pattern of life may be ethically repugnant to all advanced technological societies that reach the starfaring stage. Many aliens may have come to our solar system in the past, but they all have chosen not to disturb our world. Some of them may have set the Earth aside as a sort of wildlife preserve or scientific observation station, perhaps as a cosmic "zoo." Vastly superior intelligences may be watching and studying the Earth at this very moment. We humans do not see these beings because they do not want to be seen. Aliens may be acting as cosmic "game wardens," defending the human race against "poachers" or other unscrupulous creatures in the stars who might be tempted to take unfair advantage of an immature technological society such as ourselves. When the time is right, our guardians may choose to make themselves known. Alternatively, we may have been deemed to be a species dangerous to the rest of the galaxy and placed under strict quarantine. If so, we will not be permitted to roam very far.

The "zoo hypothesis" is by its very nature unverifiable. Such speculations necessarily lead us away from scientific fact into the realm of science fiction. Perhaps the Earth has indeed been visited in the historic past by the representatives of extraterrestrial civilizations. These creatures may have actually contacted pretechnological human societies several thousand years ago and started humankind moving along the road which was eventually to lead it toward civilization. Perhaps they are still arriving on Earth even in the present day, the frequent UFO or flying-saucer reports being evidence that our planet is under continual extraterrestrial observation. Such speculations are the subject of the next two sections.

CHARIOTS OF THE GODS?

In 1968, a Swiss author by the name of Erich von Daniken (who describes himself as an "armchair archaeologist") published a book entitled *Chariots of the Gods?*, in which he proposed that there was strong evidence that extraterrestrials had visited the Earth in prehistoric times. They came here to start humankind along the path which was ultimately to lead to the creation and maintenance of an advanced civilization on Earth. Upon arrival on Earth, one of the first acts of these superintelligent beings was to perform a number of subtle alterations in the minds and bodies of pretechnological humans by means of sophisticated genetic engineering experiments. Through the gradual processes of trial and error, new men and women were created who were more readily suited to living in an advanced technological civilization. These extraterrestrials then taught our ancestors how to construct buildings, cure illnesses, make laws, build scientific instruments, chart the skies, and perform other such tasks that are now so essential to any organized society. Although von Daniken was not the originator of this thesis, he has certainly been the most successful of its exponents, and his books have sold millions of copies worldwide. He has lectured widely throughout the world and has become an international celebrity.

Von Daniken claims that the archaeological records left behind by ancient peoples show clear evidence of the use of an advanced technology which was so far beyond the capability of anyone living at the time that they must have had outside help. According to von Daniken, the Egyptians of 4500 years ago could not possibly have built vast structures like the pyramids with such precision without the assistance of massive construction machinery and modern surveying techniques. An old iron pillar in India is said to be completely rustproof, indicating a detailed knowledge of advanced metallurgy at an early date. The erection of the massive stone statues of Easter Island is said to require engineering ability far in excess of that which the present residents are capable of. There is a museum near Baghdad in Iraq that houses several ancient jars that appear to have once been batteries. The wreckage of a Greek ship that sank at about the time of Christ contained the rusted remains of a mechanical computer. An Assyrian crystal lens from the seventh century B.C. is said to require "a sophisticated mathematical formula" for its manufacture. A Peruvian museum has several carved Inca stones depicting advanced medical techniques, such as open-heart surgery. A map originally owned by the sixteenth-century Turkish admiral Piri Re'is (dated 1513) appears to show the Earth as viewed from a point in outer space directly above Cairo. The North and South American continents can be seen, and the outline of Antarctica can supposedly be discerned, although this frigid continent was not discovered until the eighteenth century. Ancient Babylonian tablets were able to accurately record past and future eclipses. According to the Bible, the cities of Sodom and Gomorrah were destroyed by what appears to have been none other than a nuclear explosion. Many other examples are cited by von Daniken.

Von Daniken also argues that there exist ancient legends, carvings, statues, and the like that are strong evidence that there must have been some sort of contact between ordinary human beings and godlike creatures which came down from the skies. There is a large carving on a cliff overlooking the Sahara Desert that looks a lot like an astronaut dressed in a spacesuit and helmet. The lid of a Mayan tomb unearthed at Palenque has a carving of what von Daniken imagines to be an ancient astronaut operating the controls of some sort of spaceship. Levers, dials, and instruments in the cockpit can supposedly be identified, and a flaming exhaust can even be seen coming from the bottom of the craft. On the Nazca Plains of Peru there are many hundreds of miles of huge drawings of animals and other figures made by scraping aside the loose brown stony surface to expose the lighter soil beneath. According to von Daniken, these huge drawings were obviously intended to be seen only from the air. The purpose of the drawings was undoubtedly to attract the attention of extraterrestrial visitors, so that they would be induced to land their ships there. Some of the lined figures may also have served as runways or docking bays for ancient spaceships. There are legends in both Mayan and Sumerian cultures in which these societies were originally founded by mysterious beings who come down from the skies.

Von Daniken has made himself internationally known through his books and lectures, but is there any actual truth to his assertions of evidence for extraterrestrial visitations? Unfortunately, whenever any of his evidence is critically examined by informed people, his thesis falls apart. The "Great Martian God" which overlooks the Sahara (and other similar examples) turned out to be only an ordinary human being wearing a ritual mask. The "Palenque astronaut" is nothing more than an image of the Mayan king Pacal buried in the tomb underneath depicted in ceremonial dress and performing a standard ritual. An investigative team from the British Broadcasting Corporation found that the "Inca stones" depicting such modern innovations as heart transplants were in fact modern forgeries, created by natives for the benefit of gullible tourists. The destruction of Sodom and Gomorrah as depicted in the Bible does not, in fact, resemble a nuclear explosion at all. For one, the mere observation of a nuclear explosion from afar does not turn a person into a pillar of salt, the unhappy fate of Lot's wife when she turned to look back at the doomed cities. The iron pillar in Delhi, India was actually a single piece of iron smelted about A.D. 500. It is a remarkable piece of work, but it is nevertheless entirely terrestrial. The "Assyrian lens" is nothing more than a natural piece of crystal that was polished around the edges and used as an ornament. There is no evidence that it was ever intended for optical use. The "Baghdad battery" was probably a primitive electrical cell used for electroplating at about the time of Christ. A replica could produce about a half-volt of electricity for a couple of weeks, an impressive performance for two thousand years ago, but hardly evidence for extraterrestrial assistance.

The true explanation of the Nazca drawings is as yet unknown, as their builders left no written history. Several scholars think that they were a particularly odd means of keeping astronomical and historical records. It is absurd to imagine that races capable of crossing hundreds of light-years of space would require 10-kilometer runways to land once they arrived on Earth. There is absolutely no evidence for any advanced machinery having been present in the area before the current century.

A lot of nonsense has been written about the Piri Re'is map. It is a crude rendition of what early cartographers thought the newly discovered Americas should look like. None of the current rivers in South America can be identified with any certainty, and the coast of that continent meanders off to the east at the bottom of the map to join with a purely imaginary land mass. Early cartographers were often fond of adding mythical continents and islands to areas which had been poorly explored. Furthermore, a simple check of a globe of the Earth shows that it is impossible to see Antarctica from a point over Cairo. It is on the other side of the world.

The "Greek computer" is an example of mechanical gadgetry which was actually rather common at the time. Many mechanical contrivances were in everyday use throughout the ancient world, unfortunately most often strictly for such destructive purposes as warfare or else only for the private amusement of the wealthy. In Alexandria there was actually a working steam engine, although it seems never to have been employed for anything more useful than the operation of the moving parts of a temple idol to frighten the superstitious. For some reason it never seems to have occured to any of the ancients to put their machines to any sort of practical use in saving labor or in performing useful work.

Thor Heyerdahl (and others) have amply demonstrated time and time again that the mysterious Easter Island statues were indeed built by the ancestors of the present inhabitants with materials and techniques available locally. No advanced machinery or outside assistance was required. The erection of such large numbers of these statues was an essential part of a religious cult which gradually fell into decline as the small island became progressively stripped of its natural resources by an expanding population. The ecology seems never to have recovered, leaving only the relatively barren island seen today. For some reason, myths and misinformation about the Easter Island statues persist long after the truth about their origin has become well known.

Contrary to popular understanding, there is nothing mysterious about the origin and purpose of the Egyptian pyramids. They were intended as tombs for the Pharaohs of the Old Kingdom (2686-2181 B.C.) in Egypt, and a direct line of evolution to them from earlier, flat-topped tombs (known as *mastabas*) can be traced. The earliest predynastic mastabas were little more than mounds of dirt, but later examples were elaborate stepped structures made of sun-dried brick. The earliest example of what later evolved into the classic pyramid was built at Saqqara for the Pharaoh Zoser by the famous architect Imhotep. This pyramid had a steplike structure reminiscent of earlier mastabas, but it was made entirely of stone. The earliest true pyramid (with straight rather than stepped walls) was probably attempted at Meidum. The builders tried to construct the walls at an angle of 52 degrees, which appears to have had some sort of aesthetic or even magical significance (a pyramid with this angle will have a height equal to the radius of a circle of circumference equal to the perimeter of its base). Unfortunately, this angle seems to have been too steep for the construction techniques available at the time, and the outer walls of the Meidum pyramid collapsed during the latter stages of construction, leaving only a towerlike inner structure standing admidst a pile of rubble. The Bent Pyramid at Dashur was started at a steep angle of 52 degrees, but had its angle abruptly decreased near the top, probably reflecting the bitter lessons learned at Meidum. The Egyptians took no chances when they started another pyramid at Dashur; it was built at a shallower angle of 43 degrees from the beginning. Both of these pyramids were probably intended as tombs for the Pharaoh Seneferu. It was not until the reign of Cheops that the Egyptians felt sufficiently confident to try again to build pyramids with the "magic angle" of 52 degrees. The first and largest of these was the Great Pyramid at Giza, still the largest structure ever built by mankind. The Pharaohs Khephren and Mycerinus later added their own 52-degree pyramids to the complex at Giza. Shortly thereafter, the civilization of the Old Kingdom went into a period of rapid decline and eventually collapsed into chaos and anarchy. The art of the pyramid fell into decay, later editions being much smaller and so poorly constructed that they rapidly crumbled into ruins. The pyramids were all constructed by human labor, the limestone blocks being cut from nearby quarries and rolled to the site by the use of large wooden logs. They were lifted into place by the use of temporary causeways. There is no evidence of any sort of advanced machinery being used to shape the blocks or move them into place. The long history of the evolution of the pyramid is an example of an all-too-human enterprise (with learning by trial and error playing the vital role), rather than any kind of evidence for a sudden infusion of knowledge from the outside. The stone works of Egypt were indeed impressive, but they were state-of-the-art technology for the period. Any intelligent creatures capable of crossing interstellar space certainly would not have taught humans to work in stone. They would have chosen metals or plastics.

A knowledge of astronomy was common to most ancient civilizations, both in the Old World and in the New. Part of the reason for the intense interest in what went on in the skies was religious or mystical, people believing that they could somehow predict the future by observing the movements of the stars and planets. But another part was eminently practical, a knowledge of astronomy being essential for deep-water navigation and successful agriculture. One does not need to invoke the presence of alien intelligences to explain the ancient sophistication in astronomical matters. In suggesting the intervention of extraterrestrials in human affairs as the root cause of human civilization and rapid technological advance, von Daniken follows many of his contemporaries in underestimating the capabilities of primitive peoples and ancient civilizations. Our ancestors of several thousand years ago were fully as intelligent as we, and it is an insult to their memory to insinuate that they were incapable of developing a high civilization without outside help.

None of the examples that von Daniken offers as evidence for extraterrestrial visitation can stand up to any sort of critical analysis. In order to be truly convincing, he would have to produce some sort of ancient artifact, machine, or device which is unquestionably so far beyond the technological capability of the time that it could only have been produced by an extraterrestrial technology. The case for ancient astronauts would be on a much firmer footing if there was any sort of incontrovertible proof for the existence in ancient times of such things as sophisticated electronic equipment, devices constructed of modern materials like plastics or semiconductors, machinery fabricated by use of advanced metallurgical engineering or sophisticated material-processing techniques, devices manufactured to extremely high tolerances, objects which could only have been manufactured in a zero-gravity or in a high-vacuum environment, or materials with evidence for the incorporation of highly radioactive, artificial elements, such as plutonium or technetium. Von Daniken would also be a lot more credible if he could produce actual evidence that the ancients had knowledge of modern ideas and concepts in fields such as physics, chemistry, or biology. Especially valuable would be any sort of ancient manuscript alluding to such modern notions as the special theory of relativity, quantum mechanics, the quark model of matter, the existence of electromagnetic waves, the germ theory of disease, the laws of heredity, the periodic table of the elements, artificial radioactivity, or the molecular biology of the gene. Von Daniken would also make a good case if he could find some sort of ancient knowledge that could not possibly have been obtained any other way than by access to advanced technology or extraterrestrial assistance. Perhaps some solid evidence can be found that the ancients knew of the planets Uranus, Neptune, or Pluto, the two moons of Mars, the Galilean moons of Jupiter, or the rings of Saturn. An excellent case for the reality of extraterrestrial intervention in human affairs could certainly be established by the discovery of ancient but accurate maps of the "dark" side of the Moon, the planet Mars, the cloud-shrouded planet Venus, or even the moons of Jupiter or Saturn, or other such knowledge which could only have been obtained by actual space travel.

In his 1976 book *The Sirius Mystery,* Robert Temple claims to have found just such an example. He describes a tribe known as the Dogon who live in a remote region of the Republic of Mali in Africa. This tribe was the subject of intense study by the French anthropologists Marcel Griaule and Germaine Dieterlen during the years 1931 through 1952. The Dogon seem to have some rather odd legends and myths. Mixed in with a lot of prescientific myth is some rather accurate astronomy. The Dogon are aware of the rotation of the Earth, the revolution of the Earth around the Sun, the rings of Saturn, and the four moons of Jupiter. In particular, they have a myth about the star Sirius in which it is imagined to possess a superdense companion that orbits about it with a 50-year period. Sirius actually does have such a companion, a very dim white dwarf star known as Sirius B. Its existence was first proposed early in the nineteenth century by the German astronomer Friedrich Bessel to explain the wobbling motions of the brighter component. But it was not until 1862 that the American optician Alvan Clark was able to spot it while testing a new telescope, and it was not until 1955 that it was possible to photograph the tiny faint star. How could a primitive isolated tribe such as the Dogon be aware of the existence of an object that can only be seen in the largest telescopes?

According to Dogon mythology, a group of "amphibious" beings known as the Nommo were the founders of all human civilization. Similar myths actually exist in several other cultures. In particular, there are ancient Sumerian legends originating from before 4000 B.C. that claim their culture was originally started by a strange creature named Oannes, which is described as a sort of cross between a fish and a crustacean. Oannes taught humankind how to construct houses, solve geometrical problems, compile laws, farm land, and perform innumerable other tasks that are now so essential to civilization. A long succession of fabulous creatures of this sort seems to have appeared to the Sumerians over several generations, each one of them helping to further civilize mankind. They are never described as gods, but are depicted as coming down from the skies. Temple suggests that the Dogon Nommo and the Sumerian Oannes were related and that they originally came to Earth from the Sirius star system.

The Dogon myth is at first sight a very convincing piece of evidence for an ancient encounter with external intelligences, and it has intrigued many people. Several people have shown sufficient interest in the story to have critically examined the entire background of the Dogon myth, in particular Carl Sagan, the well-known author and television personality. The actual explanation is not yet known, but the most reasonable theory seems to be that the legend of the superdense companion of Sirius originally came to the Dogon from Europe rather than from extraterrestrial visitors. The Dogon have been fascinated by Sirius for many hundreds of years. They have a weak cultural link with ancient Egypt, where the star Sirius was of religious and even economic significance. The yearly rising of Sirius in the sky over Egypt happened to coincide with the annual flooding of the Nile, an event vital to the success of agriculture in such an arid region. This star subsequently became deeply imbedded in the cultural and religious life of the Dogon. The Dogon were actually not nearly so isolated from the rest of the world as Temple seems to think; there were French schools in the area as far back as 1907, and the entire region lies across a well-traveled trade route. The Dogon could have first encountered Europeans as early as the latter part of the nineteenth century. During the 1920s talk of the superdense nature of the dim companion of Sirius was widespread in popular newspapers and magazines in both Europe and America, and it is possible that some foreign travelers familiar with such stories passed through the Dogon region at this time. Because of the crucial importance of Sirius in their religion, the Dogon were quite naturally curious about the Western version of the Sirius myth, and they subsequently proceeded to graft the scientific story onto their legends. By the time that Griaule and Dieterlen entered the area, the scientific and mythical aspects of the Sirius story had probably become hopelessly intermingled.

There are some technical problems with the Dogon Sirius myth. The Dogon believe that there is a *third* star in the Sirius system that orbits about the bright companion with a period of 32 years. Some early astrometric studies seemed to indicate a few suggestive extra wobbles in the orbits of both Sirius A and B that might be evidence for the presence of a third star, but the observations have not proven to be reproducible and are now seriously doubted by most astronomers. At present there is thus no reliable evidence for the existence of a third star in the Sirius system, although of course this does not necessarily mean that none will be found in the future. If a "Sirius C" is eventually discovered, the Dogon Sirius myth will certainly take on added significance. Another problem is presented by the processes of stellar evolution which must have recently taken place within the Sirius star system. Many years ago Sirius B drifted off the main sequence and ballooned out into a red giant, increasing its luminosity perhaps a thousandfold. The dying star subsequently blew off a lot of its gaseous material into space, perhaps transferring a large percentage of it onto the normal A component. Sirius B then collapsed down to produce the white dwarf seen today. Any planet in the system would have been completely roasted or perhaps even totally vaporized during the death throes of Sirius B. No living creatures could possibly have survived. Wherever the teachers of the Dogon came from, it could not have been from the Sirius star system.

LITTLE GREEN MEN?

On June 24, 1947, Kenneth Arnold, a businessman from Boise, Idaho, was flying his private airplane near Mount Rainier in Washington state when he happpened to spot a formation of odd-looking objects flying near the mountain. He later described them to reporters as being flat and shiny, with shapes somewhat like inverted saucers. The press immediately seized upon this incident and gave it widespread coverage. This was by no means the first time that odd things had been seen in the skies, but so much publicity was generated by this and later sightings that a modern myth was created; that of the unidentified flying object (UFO) or "flying saucer," the twentieth century's equivalent of witches, demons, miracles, and supernatural magic.

Immediately after the Arnold incident there were several other spectacular sightings which created national headlines. On January 7, 1948, Captain Charles Mantell was flying his F-51 Mustang fighter plane on a routine training mission when he happened to spot an unusual object flying far above him. He radioed back to his base that he was going to climb to investigate, but before he could reach the object his airplane went out of control and crashed. Captain Mantell was killed. On July 24, 1948, a DC-3 airliner piloted by C. S. Chiles and J. B. Whitted and flying on a nighttime run from Houston to Boston reported a near-collision with a large, brightly lit, cigar-shaped object which was trailing flames and appeared to have numerous portholes along its sides. On October 1, 1948, a Lieutenant Gorman flying another F-51 Mustang on a nighttime training mission reported that he had encountered a mysterious bright light. After a brief dogfight with the pilot, it suddenly vanished just as rapidly as it had come.

In most of the reported encounters, UFOs are usually described as being some sort of aircraft; but they demonstrate flying skills far beyond the capability of any aircraft currently in existence anywhere in the world. They are capable of enormous accelerations, are able to make right-angle turns, can instantaneously reverse their directions, and can reach speeds in the atmosphere of many thousands of miles per hour. They are reportedly able to appear or disappear at will. Flying saucers are usually described as being completely noiseless; not even a sonic boom is heard when these mysterious objects are seen traveling through the air at speeds many times that of sound.

When jet fighters are sent after UFOs, the craft are easily able to avoid interception. They do not seem to exhibit any obvious external evidence for the presence of any sort of propulsion system familiar to us: no propellers, no jet engines, no wings, no air intakes, no rocket thrusters, no exhausts, and no vapor trails. The shapes of UFOs are most often described as being like disks or saucers or else somewhat like long cigars. When seen at night they are usually reported either as being self-luminous or else as having extensive patterns of bright lights attached. On a few occasions actual landings of flying saucers have been reported. However, the saucers always quickly scoot away whenever people approach, sometimes leaving odd-appearing indentations or burn patterns on the ground as evidence of their presence. It is often reported that automobile engines and electrical appliances fail or else operate in a peculiar fashion when saucers are nearby.

The rash of sightings that took place in the late 1940s, just after the Arnold incident, caused serious concern in American defense and intelligence circles. The Cold War between the West and the Soviet bloc was just beginning, and there was initially a very real fear that some of the saucer sightings might be evidence that Soviet aircraft were spying on American defense installations. If the Soviets could operate spy planes with this sort of performance capability with virtual impunity in American air space, they must have suddenly and inexplicably leaped decades or more ahead of the West in aircraft design. Others suggested that the performance capabilities of the saucers were so far beyond anything even remotely accessible to humans that the craft must come from some other world. If so, what were the intentions of their builders? Could they be any threat to American security? To try and answer some of these questions, the Air Force initiated an investigation, termed Project Sign, in 1948. It was succeeded by Project Grudge in 1949, which was in turn succeeded by Project Bluebook in 1952.

Usually, the Air Force has officially adopted a debunking view toward the saucers, dismissing most of the sightings as either deliberate hoaxes or as misidentifications of otherwise natural objects. The Arnold sighting was explained as an optical mirage or illusion which can sometimes take place in clear mountain air when the conditions are just right. Many other sightings could be accounted for by various meteorological phenomena, such as sundogs, floating ice crystals, or unusual cloud formations. Captain Mantell was killed trying to reach a Skyhook balloon, which was highly secret at the time and capable of operating at altitudes as high as 70,000 feet. Lieutenant Gorman probably had a dogfight with a lighted balloon. A lot of UFO reports were ultimately traced to unfamiliar aircraft, particularly weather balloons. During the beginning years of the space age, many spurious

UFO reports were generated by decaying earth satellites or spent rocket boosters seen reentering the atmosphere, giving off a spectacular flash of incandescence as they burned up. The Chiles-Whitted sighting was explained as a spectacular meteor which happened to pass near their airplane. Some sightings were later found to have been caused simply by birds viewed under unusual lighting conditions. The Lubbock, Texas nighttime sightings of the early 1950s were caused by newly installed city lights being reflected from flocks of plovers flying overhead. Many UFO reports were subsequently shown to have been none other than misidentifications of commonly seen astronomical objects, especially the Moon or the planet Venus. A few sightings were even shown to have been deliberate hoaxes, created by confidence tricksters for financial gain or by emotionally disturbed individuals seeking some sort of instant fame.

In 1952 there was a series of radar sightings over Washington, D.C. which momentarily chased the Korean War off the front pages of American newspapers. Jet interceptors were sent after the saucers, but they found nothing. Donald Menzel, a well-known physicist teaching at Harvard University, studied this and other similar radar sightings and concluded that they were probably produced by temperature inversions. An inversion layer is a blanket of warm air lying immediately above a layer of cooler air. Radar waves can be reflected from these layers, causing spurious responses to be received and giving radar operators the false impression that solid objects are present.

The Air Force investigations became embroiled in controversy almost as soon as they begun. Like most other defense projects of their time, the saucer investigations took place under a heavy shroud of secrecy. There was initially some justification for the extreme security. It was at first seriously feared that the Soviets might be responsible for some of the sightings. In addition, it was anticipated that the excessive amounts of press publicity sure to be generated by a completely open investigation might clog intelligence and security networks with so many saucer reports that it would never be possible to sort them all out. In retrospect, however, much of this caution was unnecessary and was undoubtedly more a result of the tendency of large bureaucratic organizations mindlessly to classify everything in sight than it was the result of any real need for secrecy. For whatever reason, the secrecy did immeasurable harm. For one, it gave many ordinary Americans the distinct impression that the Air Force knew much more about the saucers than it was telling. Donald Keyhoe, a retired Marine Corps officer, wrote a series of best-selling books that accused the Air Force of having clear and unmistakable evidence for the extraterrestrial origin of the saucers and of deliberately concealing the truth from the public. This was perhaps done out of a misguided fear of causing a mass panic or else out of a sense of acute embarrassment at being unable to do anything about the large numbers of alien craft regularly flying through American airspace. Because of the secrecy, a lot of rumors began to spread. Captain Mantell's airplane was said to have been deliberately shot down by the saucer it was chasing. The Air Force was reported to have several films showing flying saucers following along playfully after captured German V-2 rockets during launches from White Sands Proving Grounds in New Mexico. Several disappearing aircraft and ships (particularly in the so-called "Bermuda Triangle" area of the Caribbean) were blamed on UFOs, with people saying the crews and passengers had been taken into outer space for study. There were a few vague reports that a couple of saucers had crashed and that the Air Force had recovered wreckage and bodies. There was even a rumor that the Air Force had finally forced a saucer to land and had captured its crew. Sinister agencies of the United States government were reported to be actively suppressing the truth about the origin of UFOs, even resorting to sending mysterious "men in black" to silence any investigator who happened to get too close to the truth.

The number of flying saucer reports varies from year to year, generally increasing or decreasing as public interest in the subject grows or diminishes. The Arnold sighting of 1947 produced several other spectacular sightings, as we have seen. The UFO "flap" of 1952 (during which jet interceptors chased phantoms back and forth across the skies over Washington) may have been a result of the public interest in flying saucers generated by the release of the motion picture *The Day the Earth Stood Still* a few months earlier, which described a UFO landing near the Capitol building. The increased public interest in things extraterrestrial that was generated by the advent of earth satellites and space travel in the late 1950s produced as a side effect yet another wave of saucer sightings. Perhaps the most intense saucer "flap" of all took place in the mid-1960s. A truly spectacular series of sightings took place, particularly in New England. The great Northeastern power blackout of 1965 was even blamed on UFOs. Scarcely a day went by without at least one magazine article, newspaper story, or television program on the subject of flying saucers. This unprecedented wave of sightings may have been a result of the intense public interest in the planet Mars generated by the release of the first Mariner photographs of its surface. To still public criticism (and also to dampen public interest in UFOs), the Air Force established an independent commission, chaired by Edward U. Condon of the University of Colorado, to take a fresh and unbiased look at the whole UFO question. After a couple of years of study, the Condon Commission released its report. They examined some of the more spectacular sightings and concluded that they all had perfectly natural explanations which did not require the presence of any sort of extraterrestrial spacecraft. Consequently, the commission suggested that little benefit to science could be derived from continued study of UFOs and recommended that the Air Force close down Project Bluebook. This was finally done in 1969. The Condon report seems to have convinced many people that there was nothing sinister flying over their heads, and public interest in flying saucers began to decline, the number of reported sightings dropping off drastically. The late 1970s saw another resurgence of interest in UFOs (as well as more sightings), all a result of the motion picture *Close Encounters of the Third Kind,* which shows the first meeting between earthlings and extraterrestrials.

There have been a few occasional reports of "close encounters" with the crews of flying saucers. The best known of the contactees is George Adamski, who claims to have met with Venusians during a 1952 saucer landing. He later visited with creatures from Mars and Saturn and was even given rides in their saucers. Adamski authored a series of best-selling books and became an international celebrity.

In most of the encounters, extraterrestrials are reported to look much like ourselves, only with much greater physical beauty and far superior morality. They are keeping the Earth under constant surveillance, preventing us from destroying ourselves by our misuse of technology. However, when these UFO folk are reported to have vital words of wisdom for humankind, these messages invariably turn out to be no more than simple platitudes—for example, we should love each other and stop fighting so many wars. The details about how this is to be accomplished never seem to be made very clear. If extraterrestrials truly have such an all-consuming interest in our welfare, it seems at first sight rather odd that saucer crews invariably pick only relatively obscure individuals to contact. They never seem to want to meet with anyone important—no presidents, kings, prime ministers, legislators, or military officials. A few of the UFO folk are reported to be permanently living here on Earth, having infiltrated important posts in government and industry in many of the world's nations. They will intervene to save humanity whenever the end appears near. Occasionally, however, extraterrestrials are reported as being monsters or at best frightening creatures. Barney and Betty Hill reported (under hypnosis) that they had been abducted by the crew of a UFO during an automobile trip through New Hampshire in 1961 and that Betty had been dragged aboard the craft and forcibly subjected to a medical examination.

Even though most UFO reports undoubtedly do have quite natural explanations, is it nevertheless possible that a few of them actually do represent evidence that the Earth is under extraterrestrial surveillance? If the laws of physics that we have discovered are truly universal in scope, the answer must be no. The performance characteristics reported for UFOs make it impossible that they could be any sort of spacecraft. It is unrealistic to expect any material object to be able to pass through the air at speeds of several thousand miles per hour without making any sound, not even a sonic boom. Flying saucers never seem to have any sort of propulsion system which we can recognize; no propellers, jet exhausts, or rocket thrusters. Without these, there is no means by which a craft can accelerate. It is even less likely that such high-performance craft could be carrying living creatures inside. The rapid accelerations and right-angle turns routinely performed by UFOs would be instantaneously fatal for any crew members—they would be smashed into jelly on the interior walls of the ship. People who are committed to a belief in the extraterrestrial origin of the flying saucers counter such practical objections by suggesting that the UFOs must be powered instead by some sort of "antigravity" field, "etheric" force, "magnetic" propulsion, "space-warp", or other such futuristic means that are currently found only within the pages of science fiction novels. Since the laws of physics deem such propulsion schemes either highly impractical or completely impossible, their serious proposal as means for interstellar space flight is nothing less than a complete abandonment of the methodology of science and a return to a belief in supernatural magic.

If even a small fraction of the saucer reports are to be taken seriously, extraterrestrials must find the human race so fascinating that they visit us on a daily or even an hourly basis. The Earth must be the crossroads of the universe. However, the saucers never seem to accomplish anything significant when they do get here, certainly not enough to justify the cost of the trip. All they ever seem to be interested in is buzzing airplanes, chasing cars, or disrupting electrical power. On rare occasions they may actually touch down in an open field and quickly scurry away. At the very worst, they may carry off an airplane or two. Is this sort of activity worth the time and expense needed to come to Earth? Any race of beings sufficiently advanced to be able to cross interstellar space is unlikely to find our technology all that interesting. If life and intelligence have existed throughout the galaxy for a very long time, it is improbable that the human race will rate any more than a passing glance from advanced extraterrestrial civilizations. A few alien anthropologists might be expected to show up to observe yet another new technological civilization making its first steps, but it would certainly not be worth their while to fill our skies with entire fleets of starships. If we have indeed been noticed by advanced extraterrestrials, they could learn about us far more rapidly and cheaply by listening to our radio and television signals than they ever could by attempting to visit our planet in person.

The beliefs in ancient astronauts and flying saucers from outer space are in reality thinly disguised religions, with astronauts and spacemen taking the place of gods and angels. The Biblical myth of the creation of man in the image of God is replaced by von Daniken's gene-splicing ancient astronauts using the human race as laboratory test animals. Instead of the angels of the Lord, flying saucers now keep watch over the world. If the human race ever gets itself into serious trouble, aliens will step in at the eleventh hour to bail us out, much in the same way that God Himself was once expected to intervene whenever the end appeared near. The cults which have sprung up around these particular modern myths express longings for simple answers to the complex problems which confront humankind. They have filled a vacuum left by some of the more conventional religions, which many people find no longer able to provide spiritually meaningful or emotionally satisfying answers appropriate for the modern world. Unfortunately, the specific ideas and doctrines upon which these cults are founded have very little chance of corresponding to reality and can indeed be refuted in detail by people who have specialized knowledge in the fields involved. Because the question of the existence of extraterrestrial intelligence touches basic nerves in the human psyche and is so critically relevant to the future of mankind itself, it is essential that all discussions about the subject take place within the context of the known laws of science and universally verifiable facts. Any major departure from this framework could permit the debate to degenerate into the realm of religion and superstition. Informed speculation is certainly permissible within these guidelines, but those theories which do violence to the known laws of science or those for which there is no credible evidence certainly are not. This is not to argue for any sort of censorship or suppression of nonconformist ideas, but specialists within the scholarly and scientific communities have a moral obligation to publicly refute those sensational pronouncements about life and intelligence in outer space which they know to be demonstrably false. If recognized experts leave these misstatements unchallenged, the layman might be misled into believing that there is some kernel of truth in such pseudoscientific doctrines as the ancient astronaut hypothesis or the extraterrestrial origin of flying saucers. If the public dialogue about extraterrestrial intelligence is allowed to become dominated by charlatans who mouth obvious falsehoods, serious scientists and scholars who might be able to make significant contributions will be frightened away from the field, in much the same way that the controversy over the Martian canals drove many astronomers out of the field of planetary studies altogether.

If myths surrounding such topics as ancient astronauts and flying saucers become widely accepted by a significant fraction of the educated public, there is a danger that the human race might ultimately be tempted to exert something less than its maximum effort toward the solution of the critical problems which threaten its existence, much in the same spirit in which some of the early Christians sat idly by waiting for the end of the world to come. It would be pleasant to have assurance that there are superhuman intelligences out there in space who take some special interest in our well-being, and it is true that the theories of ancient astronauts and flying saucers from outer space make the universe a much more interesting place than it otherwise would be. However, the universe has no obligation to be interesting to us humans, and there is certainly no requirement that it take any degree of interest in our particular welfare. Any rational consideration of our position in the cosmos tells us that we will survive or perish upon our own efforts and our own efforts only. We must beware lest we foist our own fears and prejudices upon the rest of the cosmos. We must be courageous enough to face the universe as it really is, not as we would like it to be.

CHAPTER FIVE

THE SEARCH FOR INTELLIGENT SIGNALS

No matter what startling scientific discoveries and rapid technological advances await us in the future, it is probably safe to say that high-speed interstellar space flight will be one of the most challenging and difficult tasks that we humans have set for ourselves. The dispatching of even a minimal exploratory mission to a nearby star at a speed as slow as a tenth of the velocity of light is currently far beyond human capability and is likely to remain so for the indefinite future. The advent of large-scale interstellar transport will be practical only after we have achieved total mastery over our own solar system and have attained ready access to such energy resources as fusion technology or perhaps even antimatter fuel. Even then, a high-speed flight to a nearby star will require the commitment of a large fraction of the available resources of the entire human race. Even if there are millions of advanced technological civilizations in the galaxy, the nearest one is likely to be hundreds or even thousands of light-years distant from Earth. It will be prohibitively expensive to search for alien intelligences by actually launching exploratory missions in any way analogous to the long ocean voyages of the sixteenth century that open up the globe to the nations of Europe. However, unlike humans during the age of discovery, we now have another way to locate and establish contact with other beings like ourselves. This is via an exchange of radio signals. The first practical high-speed starship may be centuries away, but equipment is presently available on Earth which can send a radio message capable of being received and understood by another communicative civilization located anywhere in the galaxy.

The discovery of radio waves is one of the most important achievements of the nineteenth century. The existence of electromagnetic waves was first predicted by James C. Maxwell in the 1860s. He proposed that oscillating electric charges should emit traveling waves of vibrating electric and magnetic fields which propagate through space at the speed of light. Visible light, infrared, ultraviolet, and even X-rays and gamma rays are nothing more than electromagnetic waves which have different vibrational frequencies. The wavelengths and frequencies of the main subdivisions of the electromagnetic spectrum are given in Table 5.1.

OPTICAL AND RADIO TELESCOPES

The ancients knew of the heavens only through their unaided eyes, but the advent of the telescope in the early seventeenth century made possible great improvements in the human's capability for direct

TABLE 5.1 The Electromagnetic Spectrum

Type	*Frequency (Hz)*	*Wavelength (meters)*
Audio	10^1 to 10^4	10^7 to 10^4
Radio	10^4 to 10^9	10^4 to 10^0
Microwave	10^9 to 10^{11}	10^0 to 10^{-3}
Infrared	10^{11} to $4x10^{14}$	10^{-3} to $7.5x10^{-7}$
Visible	$4x10^{14}$ to $7.5x10^{14}$	$7.5\ 10^{-7}$ to $4x10^{-7}$
Ultraviolet	$7.5x10^{14}$ to 10^{17}	$4x10^{-7}$ to 10^{-9}
X-Ray	10^{17} to 10^{20}	10^{-9} to 10^{-12}
Gamma Ray	10^{20} and above	10^{-12} and below

observation of the stars and planets. Optical telescopes are basically of two types, *refracting* and *reflecting*. The refracting telescope was the first to appear. It consists of a long tube with a series of glass lenses placed inside in such a manner that a magnified image of a distant object is projected onto the human eye or onto a piece of photographic film. Contrary to popular lore, Galileo did not actually invent the telescope, but he was the first to put it to use in making astronomical studies. The first reflecting telescope was built by Isaac Newton nearly a century later. It consists of a long tube, open at one end and with a large parabolic mirror at the other. The light from a distant object enters the front of the telescope and is reflected off the mirror and brought to a focus at a point near the front of the tube. A much smaller mirror is placed near the focal point of the parabolic mirror to project the image onto an eyepiece or a sheet of film.

All telescopes, of whatever type, are rated according to their light-gathering power and their angular resolution. One would like to have a telescope with as large a light-gathering ability and as precise an angular resolution as possible. The laws of physics themselves set the limits to what can be achieved. The amount of light which enters a telescope is directly proportional to its cross-sectional area: the larger the telescope the easier it is to observe a dim object. The wave nature of light sets the limit to the best angular resolution which can be obtained. The *angular resolution* ($\delta\theta$) of a telescope is defined as the minimum angular separation between two equally bright stars which can be resolved. It is given by the formula:

$$\delta\,\theta = 1.029\ \lambda/d \qquad (5\text{–}1)$$

where d is the diameter of the telescope and λ is the wavelength of the light. The larger the telescope diameter, the smaller the angular resolution that can be achieved. The shorter the wavelength of the light used for the observation, the better the resolution. The mini-

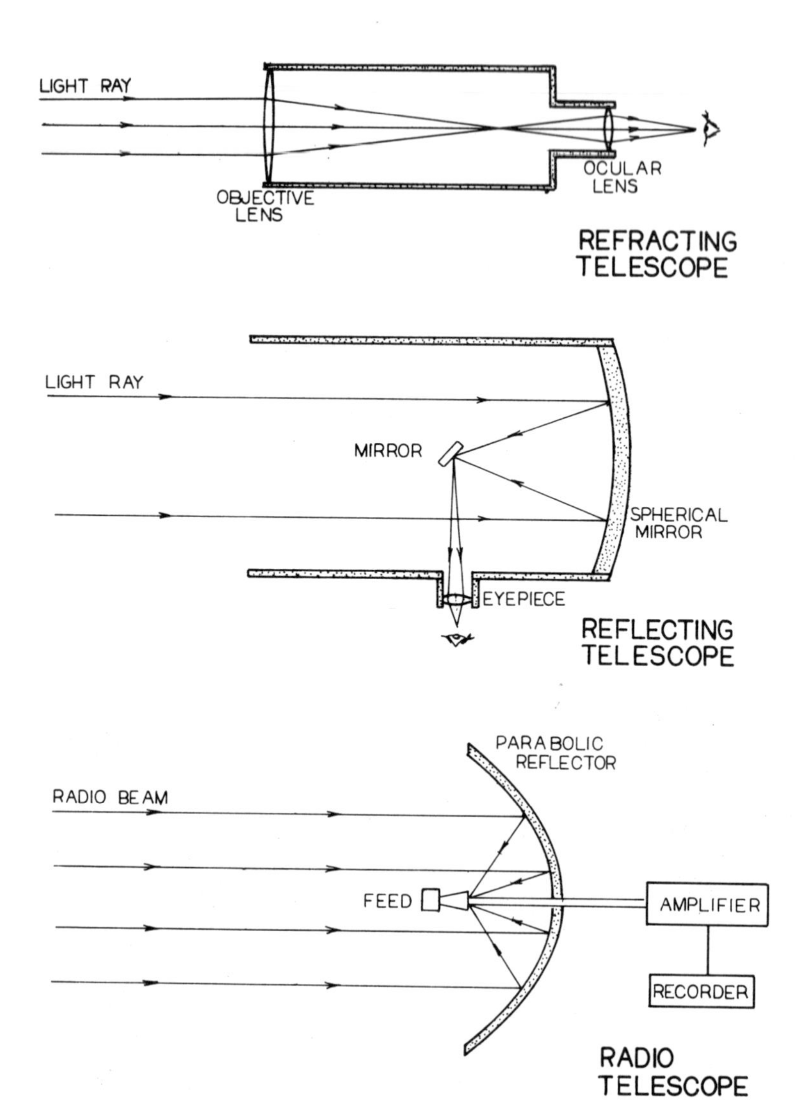

FIGURE 5.1. Optical and radio telescopes.

mum angular resolution of the 200-inch telescope on Mount Palomar in California is 0.02 second of arc (when using visible light). This would be good enough to resolve two points on the Moon only 60 meters apart. Unfortunately, this degree of precision cannot be achieved in actual practice because of the distortion produced by the turbulence in the Earth's atmosphere. The best that can usually be done is about 1 to 4 seconds.

Optical telescopes primarily use visible (and on occasion infrared and ultraviolet) light for their observations. In the years following World War II, a lot of attention has been paid to other regions of the electromagnetic spectrum, particularly the radio band. Radio astronomy came into being purely by accident during the early 1930s when Karl Jansky, of the Bell Telephone Laboratories, was trying to track down sources of noise that were interfering with short-wave radio communications. He noticed that his equipment was picking up a certain noise pattern at about the same time every day, and he later determined that the source of the noise was actually the Milky Way. After the war ended, new innovations in radar and electronics made it possible to build sensitive receivers which could explore this new region of the spectrum in greater detail. A radio receiver specially adapted for astronomical work has come to be known as a *radio telescope.* It operates on exactly the same principle as a reflecting optical telescope. There is a large parabolic metal dish that reflects incoming parallel radio waves and focuses them onto a feed mechanism above the center. The collected radio signal is then sent into an electronic system where it is amplified and recorded. By

FIGURE 5.2. Jodrell Bank radio telescope, located near Manchester in England.

making a few simple modifications, a radio telescope can be quite readily converted into a radio transmitter. This is done by replacing the feed mechanism at the parabolic dish focus with an intense source of radio waves. These waves are directed toward the parabolic dish, where they are reflected and formed into a virtually parallel beam of radio waves.

Since the radio signals impinging upon the Earth are incredibly weak, the parabolic dish of a radio telescope must be of appreciable size in order to pick up a measurable signal. The first large, fully steerable parabolic radio telescope was constructed at Jodrell Bank near Manchester in England in the 1950s. It is 60 meters in diameter. By rotating the entire dish one can aim it at any point in the sky. The largest fully steerable radio telescope currently in operation is 100 meters in diameter and is located at Effelsburg near Bonn, West Germany. The cost and complexity of such steerable radio telescopes are considerable. The principal problem is the large size and weight of the giant dish, as well as its tendency to flex and change shape under the influence of the wind and the weather. An alternative approach is taken by the 1000-foot (305-meter) radio telescope at Arecibo, Puerto Rico. A natural depression in the hilly terrain was shaped into a parabolic surface 300 meters across and covered with steel mesh to act as a reflector. The feed system is suspended by cables above the center of the dish. The dish itself is fixed, so only a very small region of sky vertically above the facility can be observed. However, the feed system can be moved back and forth by a small amount so that points as far as 20 degrees away from the vertical can be studied. The rotation of the Earth makes it possible for most points in the sky to be observed for short times each day.

Just like optical telescopes, radio telescopes are rated according to their signal-gathering ability and their angular resolution. The angular resolution of a radio telescope is also given by Equation 5.1, the principal difference being that the wavelength of the detected radiation is so much longer. The angular resolution of the Arecibo radio telescope is 2 minutes of arc when it is receiving microwaves of 21 centimeters in wavelength. This is substantially poorer performance than is possible with even relatively small optical telescopes. Because of the longer wavelength of radio waves as compared to the wavelength of visible light, radio telescopes would have to be many kilometers in diameter before they could begin to compete with the resolving ability of optical telescopes. One way to overcome this problem is to use two or more radio telescopes separated by large distances and point them all in the same direction. The signals received by each of the individual telescopes are sent to a common location, where they are electronically combined. This is the strategy followed by Very Long Baseline Interferometry (VLBI), where the individual radio telescopes are separated by intercontinental distances of thousands of kilometers. VLBI is thus capable of simulating the resolving power of a radio telescope almost as large as the Earth itself. It can achieve angular resolutions as small as 0.02 second of arc, significantly better performance than is possible with the largest optical telescopes.

THE WATER HOLE

If many other intelligent communicative civilizations exist in the galaxy, some of the electromagnetic signals which they emit must currently be impinging upon the Earth. Most of these signals may come from the unintentional but unavoidable "leakage" of radiant energy that would be a natural by-product of the routine, everyday activities of advanced extraterrestrial societies. It may be impossible to distinguish these signals from purely natural background "noise." On the other hand, a few signals may actually have been beamed toward Earth as part of a conscious effort by an alien civilization to make its presence known. Because of the extreme distances, any intelligently produced signal is certain to be extremely weak. We have no idea *a priori* where these signals will be coming from, so we have no way of deciding in which direction to point our receivers. In addition, we are initially uncertain of the frequency which extraterrestrials will employ for the transmission of their signals. Any choice of listening frequency that we ultimately make will depend on our assumptions about the types of alien technological societies that are present in the galaxy, as well as on our analysis of the psychological motives of intelligent creatures about which we know absolutely nothing. However, there are means of making intelligent choices of the regions of the electromagnetic spectrum it will be most profitable to search for intelligently produced signals. Such a choice is forced upon us by the nature of the Earth's atmosphere as well as the characteristics of the various types of interfering noise sources that are present.

The Atmosphere

Any ground-based equipment listening for signals from outer space is restricted to those frequency bands over which the Earth's atmosphere is transparent. This requirement limits our options quite severely. The absorption spectrum of the Earth's atmosphere is given in Figure 5.3. Audio- and radio-frequency signals below 1 to 10 MHz

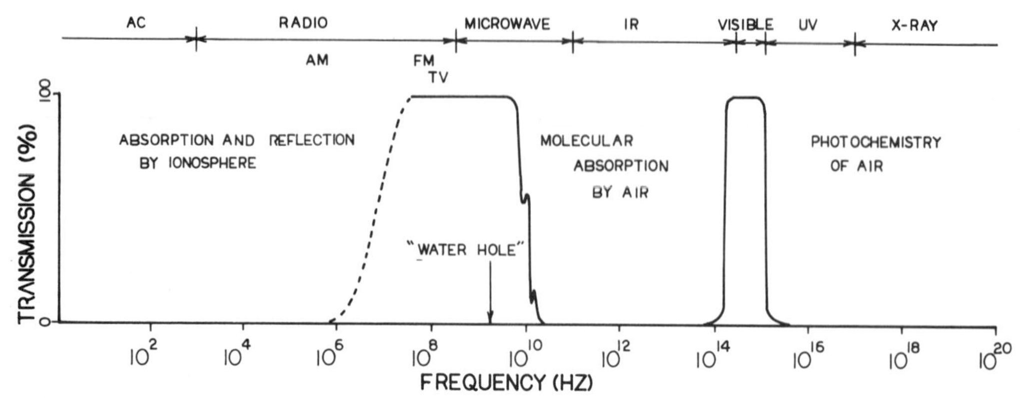

FIGURE 5.3. Absorption spectrum of Earth's atmosphere.

are strongly absorbed and reflected by the Earth's ionosphere. This includes the entire AM radio band. There is another strong absorption band between 10^{10} and 10^{14} Hz in frequency, which includes the upper part of the microwave as well as all but the very highest frequencies of the infrared region of the spectrum. This absorption is caused by the excitation of rotational and vibrational motion in air molecules as radiation in this band passes through the atmosphere. There is a third absorption band extending above 10^{15} Hz in frequency, which includes the upper part of the ultraviolet band as well as the entire X-ray and gamma-ray regions of the spectrum. This strong absorption is produced by photoionization of air molecules by the highly energetic photons which make up this short-wavelength radiation. The Earth's atmosphere (as well as the atmosphere of any other earthlike planet) is therefore completely opaque to electromagnetic radiation over most of the frequency bands in the electromagnetic spectrum. There are, however, two relatively narrow "holes" in the electromagnetic spectrum where the atmosphere is largely transparent, one in the radio and microwave region (10^7 to 10^{10} Hz) and another in and around the visible region (10^{14} to 10^{15} Hz). Earth-based telescopes capable of detecting signals from outer space are thus limited to one or the other of these two frequency bands. Other bands would be accessible to us only if we constructed our receivers in outer space above the atmosphere.

Which of the two atmospheric "windows" would give a better choice for interstellar communication, the visible or the microwave? The spectrum of the background noise may provide the answer. Noise sources will be of two distinct types: one an extrinsic incoherent and random static "hiss" that impinges upon a receiver from virtually all parts of the sky, and the other an intrinsic interference which is actually produced within a receiver's detection system itself. Sources of noise of either origin will interfere with the reception of an intelligent signal. Alien societies wishing to attract attention will undoubtedly choose to transmit their "call signals" in a region of the spectrum where the background noise is low, maximizing the probability that their signal can be detected.

The Noise of the Cosmos

Unwanted interference is most conveniently discussed in terms of a "noise temperature." This is not any sort of temperature which can be measured by a thermometer, but a rather abstract quantity related to the relative amount of power that is produced by random noise interference within a given frequency band. The noise temperature spectrum is shown in Figure 5.4. There are two important sources of externally produced interference which impinge on the Earth from the sky, the galactic synchrotron radiation and the cosmic "big bang" radiation.

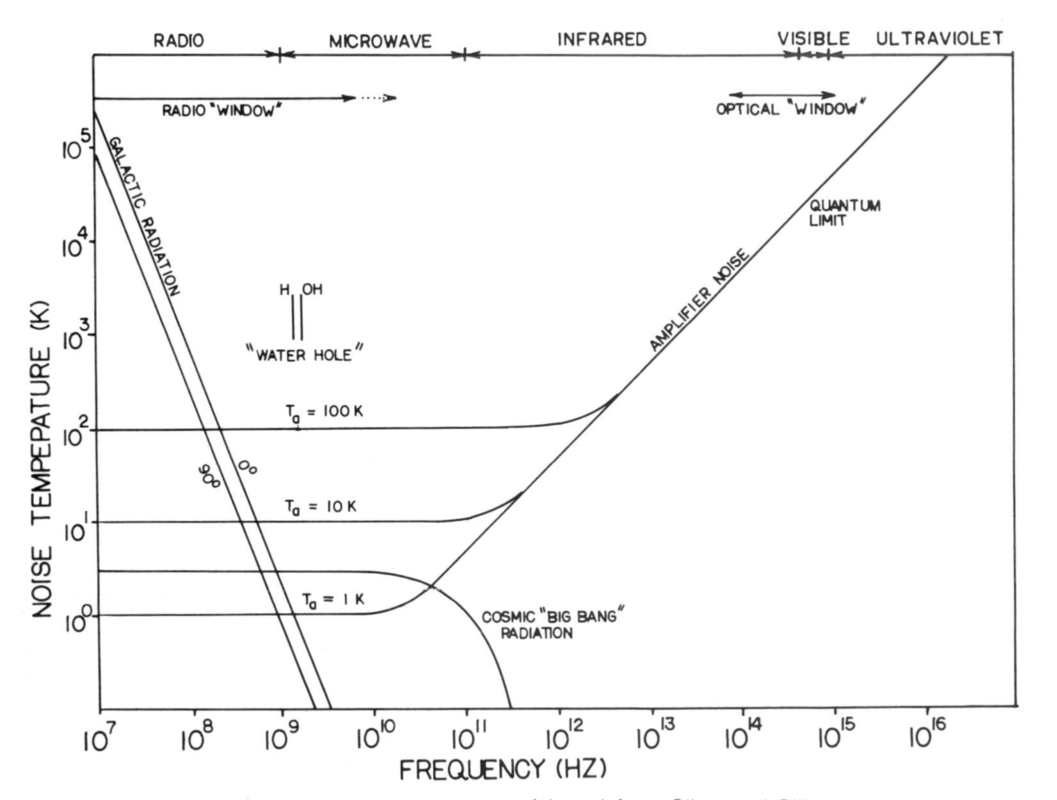

FIGURE 5.4. Noise temperature spectrum. Adapted from Oliver and Billingham (1973). National Aeronautics and Space Administration.

The galactic synchrotron radiation was the radio noise first noted by Karl Jansky in his pioneering radio astronomy studies in 1931. It comes from the entire oblate spheroidal disk of the galaxy, but is about ten times stronger when the receiver is pointed within the galactic plane than when it is at right angles to it. Most of the galactic noise is confined to the regions of the spectrum below 1 GHz in frequency, with the interference getting progressively worse the lower the frequency becomes. It originates in the weak magnetic field which permeates the entire galaxy. Highly energetic electrons emitted during past supernovae and other such violent events are spiralling around these field lines and emitting strong synchrotron radiation in the process.

The cosmic "big bang" radiation was discovered quite by accident by Arno Penzias and Robert Wilson of the Bell Telephone Laboratories while they were checking a horn-shaped microwave antenna for extraneous noise sources. They happened to be operating at a frequency of 4.28 GHz, and after all possible terrestrial microwave interference had been ruled out there was still a residual signal which could not be eliminated. At the same time, workers at Princeton University had predicted that the radiation emitted at the time of the creation of the universe many billions of years ago should still be observable as a very weak background microwave emission coming from all parts of the sky. The noise signal picked up by Penzias and Wilson is apparently the radiation left over from this incredibly violent event at the beginning of time. Subsequent measurements by other workers using both ground-based and airborne equipment have shown that the shape of the emission spectrum closely matches that of a theoretical "blackbody" radiator that has an effective temperature of about three degrees above absolute zero. The cosmic "big bang" radiation is the dominant sky noise source over most of the microwave and near-infrared regions of the spectrum.

All amplification and detection systems invariably have intrinsic internal sources of noise that are produced by random thermal fluctuations as well as by the quantum, particlelike nature of the radiation being detected. The quality of a detection system is specified by its "effective temperature" T_a. The lower the effective temperature, the better the receiver and the smaller the amount of intrinsic noise originating from within the system. Over the radio and microwave bands the noise temperature of a detection system is approximately independent of frequency and is numerically equal to the actual thermal temperature at which the equipment is operating. It is therefore possible to reduce substantially the deleterious effects of thermal noise on radio or microwave detector performance simply by cooling the system down to cryogenic temperatures by bathing it in liquid nitrogen or liquid helium. Microwave amplifiers with noise temperatures as low as a few degrees Kelvin have actually been built, and as cryogenic techniques continue to improve further significant advances may be expected. At higher frequencies, however, one eventually runs up against the quantum nature of the detected radiation itself as the limiting factor. The signal being detected actually consists of a large number of photons. The energy of each photon in a beam of radiation is incredibly small. There are so many photons in a beam of light or a radio wave that quantum effects are not usually noticed in everyday life. However, the energy of a photon is proportional to the frequency of the radiation. As the frequency is increased, a signal of given power has progressively fewer and fewer photons. Since the minimum possible fluctuation in the energy of a beam of radiation is the energy of a single photon, the relative amount of noise power in a signal will increase as the frequency increases. Because of this quantum effect, at frequencies above 10^{11} to 10^{12} Hz the noise temperature of a detection system becomes significantly larger than its effective thermal temperature. The amount of noise gets progressively worse and worse as one increases the operating frequency into the visible and ultraviolet bands of the spectrum. No amount of clever engineering can eliminate this sort of interference, as it arises from the quantum nature of the universe itself. As a result, detectors operating in the visible or ultraviolet regions of the spectrum will always have a poorer noise performance than those operating in the microwave.

Note that there is a broad minimum in the overall noise spectrum (from both extrinsic and intrinsic noise sources) in the frequency band between 10^9 and 10^{11} Hz. This includes virtually the entire microwave region of the spectrum. Any intelligent species interested in interstellar communication with the minimum noise interference possible will be forced to focus their attention upon this particular frequency band. All communicative societies that have at least attained Type I status will be aware of the particularly attractive features of the microwave band for establishing contact with others like themselves. Fortuitously, this ideal communication band happens to coincide almost perfectly with the microwave atmospheric "window" discussed earlier. We can search for intelligent extraterrestrial signals from the ground.

The Magic Frequency

Even though the microwave band is the most profitable region of the spectrum to search for the presence of intelligently produced signals, the band itself is nonetheless rather broad. Must all of it be searched in order to have a reasonable chance at finding an intelligent signal? In 1959, Giuseppe Cocconi and Philip Morrison published a scientific paper in the prestigious British journal *Nature,* in which they argued that there exists a unique objective standard frequency lying in the microwave band that should be known to all communicative technological civilizations in the galaxy. This is the frequency of the radio emission of hydrogen atoms, the most abundant element in the universe. The tenuous interstellar atomic hydrogen emits a steady background noise at a precise frequency of 1420 MHz (or a wavelength of 21 centimeters). Because of its importance to radio astronomy, the narrow frequency band in and around 1420 MHz is protected by international agreement against interference by military or civilian broadcasters. Every civilization with access to radio astronomy will likewise be aware of the special significance of the hydrogen line. Many civilizations around other stars may at this very moment be carrying out extensive astronomical studies of this radio emission. Anyone wishing to draw attention to himself would be wise to transmit his call signal at this same frequency.

However, since 1959 it has been known that the neutral hydrogen emission line is not all that unique and that there are radio emissions from many other interstellar molecules and atoms. In particular, there is a series of four microwave lines starting at 1662 MHz emitted by the hydroxyl radical (OH). In 1971, a group of scientists and engineers gathered in California under NASA sponsorship to evaluate various alternatives for SETI (Search for ExtraTerrestrial Intelligence) projects then under consideration. The team called itself "Project Cyclops" and paid particular attention to the best choice of frequency band in which to search for intelligent signals. The Cyclops team proposed that the relatively narrow spectral region between the hydrogen and hydroxyl emissions would be a particularly profitable band in which to look for intelligent radio signals. Since H and OH are the dissociation products of water, this band has come to be known colloquially as the "water hole." The words of the final Cyclops report are worth quoting here:

> Surely the band lying between the resonances of the dissociation products of water is ideally suited and an uncannily poetic place for water-based life to seek its kind. Where shall we meet? At the water hole, of course!

Bandwidth

Once we decide on a listening frequency, what value do we choose for the receiver bandwidth? Every receiving system has a finite bandwidth; that is, it allows only those radio waves with frequencies lying within a certain range (termed the bandwidth or bandpass) to pass through the detection system and be amplified, excluding all others. In this way an AM radio receiver can tune to one specific station, ignoring all the others. The choice of ideal receiver bandwidth for use in searching for intelligently produced microwave signals will ultimately depend upon the assumptions that are made about the types of signals that extraterrestrials are likely to send toward Earth. This is

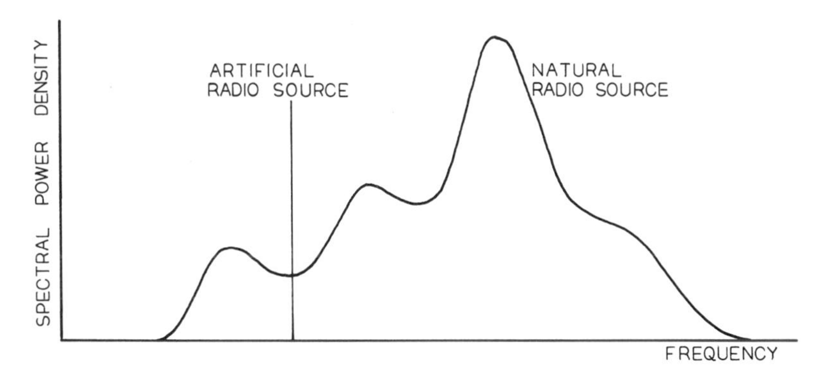

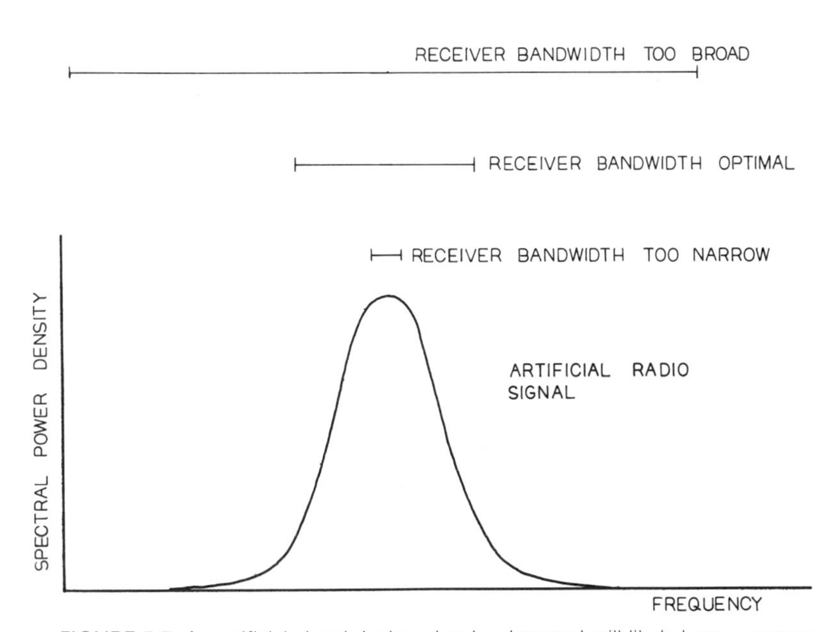

FIGURE 5.5. An artificial signal designed to be detected will likely have a narrow bandwidth as compared to natural radio sources. The most efficient detection system would have an input bandwidth which exactly matches the bandwidth of the transmitted signal.

difficult to decide in advance, but it is probably safe to say that any call signal specifically designed to attract attention will be transmitted in such a way as to be easily detected. Furthermore, such a signal will almost certainly be designed so that it will be instantly recognized by any observers as having none other than an intelligent origin. Most natural radio and microwave sources give out a broadband pattern of noise that is spread out over many MHz in frequency. One sure way for an extraterrestrial civilization to attract attention would be to transmit a signal having as narrow a bandwidth as feasible, perhaps only a few Hz in width or even less. Such a signal should stand out like a sore thumb from the background noise of the rest of the galaxy and would alert any observers that here was something worthy of more attention. So we would expect intelligent signals that are specially designed to attract our attention to be transmitted at relatively narrow bandwidths. The most efficient arrangement possible would be for the intelligent signal and the receiver to have exactly the same bandwidth. If the receiver bandwidth is appreciably narrower than the intelligent signal, some of that signal will be unintentionally blocked out. If the receiver is set at a bandpass much broader than the signal, a lot more sky noise will be picked up without any increase in signal.

Integrating Time

Another factor to be considered is the length of time that a receiving antenna should be pointed toward a likely source. How long should we listen before we give up? The chances of picking out a weak monochromatic signal from the cosmic background noise can be markedly increased if the receiver electronically averages the incoming signal over a long time period. During the averaging process, a coherent monochromatic signal will tend to add, whereas the incoherent background noise will tend to average to zero. The longer the averaging time, the better the chance of detecting a weak signal. The time spent on electronic averaging is sometimes called the "integrating time." Unfortunately, the longer the time spent on any one source, the less the time available for all of the others. The use of excessively long integration times will make the search for intelligent signals very lengthy. Within a thousand light-years of Earth there are one to three million sunlike stars, all of which will have to be searched. If a single day is spent on each star, it will take ten thousand years to finish the project. Some sort of tradeoff will obviously have to be made between the length of time spent on each star and the length of time allocated for the entire search.

OZMA AND ITS SUCCESSORS

Ozma the Pioneer

The first known attempt to listen for intelligent signals from other stars was made by Frank Drake, then working at the National Radio Astronomy Observatory (NRAO) at Green Bank, West Virginia. As an undergraduate student at Cornell University, he had become interested in the question of intelligent life on other worlds. During a three-year hitch in the Navy he worked with sophisticated electronic equipment aboard a heavy cruiser. After leaving the Navy, he received a doctorate in astronomy from Harvard University. For a short time thereafter he worked for Harold Ewen, one of the scientists who discovered the 21-centimeter hydrogen emission.

In 1958, Frank Drake joined the staff of the newly established NRAO. It had been built in a remote area of West Virginia, hopefully well away from any terrestrial radio interference. One of the top priorities of the new facility was the observation of the 21-centimeter microwave emission coming from the neutral hydrogen scattered throughout the galaxy. Drake proposed to point NRAO's new, 85-foot radio telescope toward a few nearby stars and listen for evidence of any intelligent signals being transmitted at frequencies near that of neutral hydrogen. He gave his proposal the name "Project Ozma," after the princess of the mythical land of Oz depicted in the fictional works of L. Frank Baum. He chose as his first targets the two sunlike stars Epsilon Eridani and Tau Ceti. Epsilon Eridani is a class K2 star, 10.7 light-years distant in the constellation Eridanus (The River), whereas Tau Ceti is a class G8 star located 11.9 light-years from Earth in the constellation Cetus (The Whale). Both stars are in a region of the sky easily accessible to West Virginia.

On April 8, 1960, the radio telescope was turned toward the star Tau Ceti for the first time. It was observed for several hours, but no unusual signals were noted. The radio telescope was then turned toward Epsilon Eridani. Almost immediately, a strong signal consisting of eight pulses per second was picked up! Pandemonium broke out in the control room, but there was absolutely no thought of any announcement being made to the press. The Ozma team had to be careful not to jump to premature conclusions. As a check, the radio telescope was turned away from the star. The signal dutifully vanished. However, it did not return when the radio telescope was turned back toward the target. Apparently, terrestrial interference was responsible, and a later investigation showed that the pulses that had been picked up were probably produced by a secret military experiment in radar countermeasures. For the next couple of months both stars were repeatedly scanned, but nothing further was noticed. The total observation time was about 150 hours, after which the search was suspended because of the pressing needs of other experiments for telescope time. Even today some sensationalist publications occasionally print stories which claim that Ozma had actually detected intelligent signals from another world, but that some fiendish government agency is suppressing the truth.

A radio transmitter located on either one of these stars and broadcasting equally toward all directions of the sky would have required a power of at least 100 terawatts in order to have been detected by Ozma's receivers. Power levels of this magnitude have not yet been achieved on Earth. However, if the inhabitants of either star had happened to be using an exact duplicate of the 85-foot radio telescope at NRAO to beam a directional signal to Earth at the precise time that Frank Drake was listening, they could have attracted his attention with only 700 megawatts of continuous power. This power level can routinely be attained by many of the more powerful radio and radar systems currently in use here on Earth. If the inhabitants of Epsilon Eridani or Tau Ceti had attained a technological level at least equivalent to that of present-day Earth, they could easily have attracted our attention had they wished to. Either these particular stars lack communicative civilizations altogether or else their inhabitants are uninterested in making contact with us at the present time. Or perhaps they did not happen to be transmitting during the brief time that we were listening.

Later Studies

Although the results of the search were negative, the Ozma project captured the popular imagination and inspired others to carry on the effort. In 1968, Vsevolod Troitsky, of the Gorky State University in the Soviet Union, made a search of eleven sunlike stars located within 62 light-years of Earth by using a 15-meter radio telescope. In addition, the entire Andromeda galaxy was scanned (covering at least a trillion stars at one blow), although the chance of an intelligent signal being received at such a large distance (over 2 million light-years) is obviously rather small. Troitsky's team used a new type of microwave receiving system, one which was capable of breaking up the spectrum of the incoming signal into 25 separate bands each 13 Hz wide. Each of these bands was simultaneously analyzed by the electronic equipment, making it possible to search a rather broad spectral band at extremely high sensitivity in a relatively short amount of time. The receiving system was slowly scanned across a 2 megahertz-wide frequency band, taking about ten minutes to make a full pass. The frequency used by Troitsky was 0.927 GHz rather than hydrogen frequency of 1.42 GHz because of the characteristics of the available equipment, although the search was otherwise similar to that of Ozma. Nothing unusual was found.

Troitsky and his colleagues tried a different tactic in 1970. A network of omnidirectional dipole antennas scattered over a large area of the Soviet Union searched for coincidental electromagnetic pulses from the entire sky in the decimeter band (1.875, 1.0, and 0.6 GHz), without attempting to preselect any particular direction. Some of these pulses could be stray emissions from unimaginably vast astroengineering activities of Type III supercivilizations located at the far reaches of the Milky Way or in other galaxies. The number of coincidental pulses detected was apparently larger than could be attributed to pure chance, but there were more events during the day than there were at night. Troitsky cautiously attributed these coincidental pulses to solar activity rather than to extraterrestrial intelligence. This effort is apparently continuing. There have been occasional reports in the Soviet press of sequences of regular pulses being picked up by this network of antennae. Most have been subsequently traced to solar emissions, to stray terrestrial radio beams bounced off Earth's ionosphere, or to signals transmitted by earth satellites or space probes. In 1973, a sequence of pulses was detected which appeared to have no such explanation. Nikolai Kardashev suggested that these pulses could have come from a Bracewell probe located somewhere in our solar system. It is difficult to evaluate such reports because of the lack of detailed information, but it is most likely that these signals were produced by secret military activity on or near the Earth rather than by extraterrestrial intelligence. Secret Soviet or American reconnaissance satellites could have been the culprits.

In 1972, there was a renewed American interest in SETI. Gerrit Verschuur of the NRAO took a quick look at ten nearby stars (including Epsilon Eridani and Tau Ceti) during a conventional study of the hydrogen emission from the galaxy. He was using both the 140-foot and the 300-foot radio telescopes at the Green Bank facility. No unusual signals were found, although Verschuur's receiver was probably at least a hundred times more sensitive than that of Ozma. A more extensive search was made by Benjamin Zuckerman of the University of Maryland and Patrick Palmer of the University of Chicago a couple of years later with the same pair of radio telescopes. They called their study "Ozma II," and they eventually examined nearly 700 stars within 80 light-years of Earth. Some binary stars were included in the search. Each star was observed for approximately 4 minutes per day for 7 consecutive days. Zuckerman and Palmer used a frequency analyzer capable of breaking up the incoming signal into 384 individual channels, each 4 kHz wide. The received signals were recorded on magnetic tape for later analysis "off-line." A subsequent check of the recordings actually turned up some unexplained "glitches" in the signals from a few of the stars that were searched, suggesting that some further examination might be warranted. After a delay of a year, ten of the stars that had showed "glitches" were reobserved. In only one case was a second glitch found.

In 1977, Frank Drake (now at Cornell University) and Mark Stull of NASA-Ames used the giant Arecibo radio telescope to take a quick look at ten conveniently located stars during gaps in the conventional research program of the installation. The broadband signals that were received were recorded on magnetic tape and then transferred to photographic film where they were analyzed with an optical processor to see if there was any order present that might be evidence for intelligent transmission. Nothing unusual was found.

In 1976, four workers at NRAO (Jeffrey Cuzzi, Thomas Clark, Jill Tarter, and David Black) used a special narrowband receiver (capable of a few Hertz resolution), originally designed for use as part of a VLBI arrangement, to check some of the stars that had shown interesting "glitches" during the Ozma II search. Some other likely targets were also studied. Nothing worthy of note was found. In 1977, this same team of workers used five days of observing time on the 300-foot Green Bank radio telescope to listen to 200 sunlike stars. The team chose a frequency band near the OH-side of the "water hole" for the search, and they recorded the incoming signals on magnetic tape and used computer analysis for data reduction after the search was completed. They employed an advanced signal processor capable of resolving the incoming signal into 65,536 separate channels, each 5 Hz in width. A few "interesting" signals were noted on the tapes which were in all probability caused by terrestrial interference.

In 1979, John Billingham of NASA-Ames used the giant Arecibo radio telescope to study 250 sunlike stars within 100 light-years of the Earth. He used a receiver having a 4-MHz input bandpass multiplexed into a million separate channels. He swept his receiver over the entire width of the "water hole." Although he spent approximately ten minutes listening to each star, he found nothing suggestive of any sort of intelligent signal.

At Ohio State University, Robert Dixon and his staff have been trying a different strategy. They have access to a transit radio telescope which can periodically sweep across the entire sky in search of radio signals without preselecting any specific targets. The facility might be able to spot a highly powerful beacon transmitter located in our immediate vicinity. The facility began the search in 1973, using a frequency band 350-kHz wide centered on the hydrogen line. During the first few months of operation many strong, pulsed signals were picked up. These were not found in subsequent sweeps of the sky, and all are believed to be of terrestrial origin. Radar pulses from military and civilian aviation installations were the chief culprits. More advanced equipment was eventually provided by NRAO as well as by private donors, and the Ohio State team has made broad, continuous sweeps of the entire sky at irregular intervals during the past few years, using a frequency of 1.42 GHz. Only one signal even remotely suggestive of an extraterrestrial origin was ever detected. It was an intense series of pulses received in 1977, apparently coming from a region of sky where there are no nearby sunlike stars. Its origin is unknown, and it did not reappear when the same region of sky was studied again. A space probe or secret military satellite may have been the cause. The Ohio State search continues at the present day on a sporadic basis. No government funding has been provided, so the facility has to operate on less than a shoestring. The radio telescope itself is unfortunately now in danger of being torn down and replaced by a golf course.

Yet another approach has been followed by C. Stuart Bowyer and his group of graduate students at the University of California at Berkeley. They have attached their instrumentation as a parasitic experiment onto a radio telescope being used for conventional radio astronomy research. The sky locations and frequencies examined are dictated by the primary observation program, and any detection would be the result of pure chance. They somewhat whimsically gave their project the name SERENDIP, which is an acronym standing for "Search for Extraterrestrial Radio Emission from Nearby Developed Intelligent Populations." Their instrumentation has gone through several changes over the years; the most recent version is a 100-channel analyzer capable of a 500-Hz resolution over a 20-megahertz bandpass. A microprocessor attached to the detection system is programmed to search the radio waves intercepted by the telescope for any narrowband signals exceeding a preset threshold level. The data is stored on magnetic tape and is analyzed "off-line" by a computer at a later time. Various different systems have been attached to the 85-foot Hat Creek radio telescope since 1972, and periodic checks of the recordings have been made. Over the years, several thousand narrowband signals have been detected. The vast majority have been subsequently traced either to sources within the detection system itself or to terrestrial interference. A few signals may warrant further examination, but nothing definitely indicative of an intelligent extraterrestrial origin has yet been detected.

The Canadian workers Alan Bridle and Paul Feldman initiated a SETI effort in 1974, using the 150-foot radio telescope at the Algonquin Radio Observatory. They feel that the water line at 22.2 GHz is a better choice for the observation than the hydrogen line, and they initially planned to study about 500 sunlike stars. Their receiver is capable of covering a total bandwidth of 10 MHz, with a 30 kHz resolution. They have so far spent approximately 140 hours searching seventy stars within forty-five light-years of Earth. No results have yet been published. The program is now in limbo, but may eventually be resumed if and when a better receiver becomes available.

Carl Sagan and Frank Drake are collaborating on a small-scale SETI effort using the Arecibo radio telescope in Puerto Rico. Instead of examining nearby stars, they have been searching entire galaxies. Perhaps some other galaxies have produced sophisticated Type III civilizations capable of making their presence known over such vast distances. They are using a 1008-channel correlator capable of covering a bandwidth of 3 MHz with a 1-kHz resolution. So far, they have spent approximately 100 hours searching five nearby galaxies at frequencies of 1.42, 1.653, and 2.38 GHz. Nothing of interest has yet been found.

During the late 1970s the Westerbork Synthesis Radio Telescope in the Netherlands (which consist of fourteen 25-meter parabolic receivers placed in an array three kilometers long) was employed to make high–spatial resolution radio maps of the accessible sky at frequencies near the 21-centimeter hydrogen line. These maps were subsequently checked to see if there were any anomalous radio sources coincident with the positions of nearby stars that could conceivably have been produced by alien intelligences. The particular advantage of this search technique is that it does not interfere with the conventional research program of the radio telescope. Five hundred forty-two stellar positions (as well as the galactic center) were checked for the presence of anomalous emissions, but no evidence for intelligent extraterrestrial radio signals was found.

Perhaps the most sensitive SETI effort to date has been made by Paul Horowitz of Harvard University. In 1978, he used the Arecibo radio telescope to search 200 nearby stars at a frequency of 1.42 GHz for signs of intelligent signals. He employed a multichannel receiver capable of resolving the incoming signal into 65,536 individual "bins," each only 0.015 Hz wide. This sort of frequency resolution should be sufficiently sensitive to spot an omnidirectional radio beacon as weak as 100 megawatts located on a star 20 light-years distant. Horowitz assumed that any society attempting to signal Earth had accurately measured the radial velocity of the Sun and would trasmit a signal whose frequency would be received here on

Earth at the hydrogen line rest frequency of the Sun. In order to correct for the Doppler shifts of the incoming signal resulting from the Earth's orbital motion, Horowitz had to continually sweep his observing frequency during the few hundred seconds of observations of each target star. The incoming data was recorded on magnetic tape and analyzed "off-line." Horowitz looked at two hundred nearby F, G, and K stars, but excluded known binaries. The total observing time was 80 hours, but absolutely nothing was found, not even a false alarm.

The various SETI efforts so far reported are summarized in Table 5.2. All without exception have turned up negative results. This should not necessarily be interpreted as meaning that intelligent life is rare or nonexistent in the galaxy. In all, scarcely a thousand likely stars have been examined in any detail for signs of intelligently produced signals. Of these stars only a couple of dozen can statistically be expected to possess habitable planets in the first place, to say nothing of technological civilizations capable of radio communication. Obviously, many more stars will have to be searched before there is much prospect of success.

THE POLITICS OF SETI

Cyclops

The National Aeronautics and Space Administration (NASA) has periodically sought out new ideas for SETI projects. Several extremely ambitious proposals have been put forth. The best-known of these is the proposal of the Cyclops team. This consists of a circular array of over 2500 large (100-meter), fully steerable individual radio telescopes placed in a circular, orchardlike pattern over fifteen kilometers across. During the search each of the antennae will be pointed toward the same direction in the sky, and the phases and delays of the signals picked up by each will be electronically adjusted so that the amplitudes of only those signals coming from the desired direction will add when they are fed into a central receiving station. The angular resolution of the entire facility will be about three seconds of arc, which will rival the performance of some of the largest Earth-based optical telescopes. Parallel to the construction of the large array of radio telescopes is the development of a sophisticated computer system to analyze the incoming signals at the instant they are received for any evidence of intelligent origin. Since the data analysis is done in "real time," a target star could be immediately reexamined in much greater detail if its signal showed any interesting

TABLE 5.2 Summary of Searches for Extraterrestrial Intelligence

Investigators	*Observatory*	*Date*	*Antenna Radius (m)*	*Freq. (GHz)*	*Band Width (Hz)*	*Det. Temp (K)*	*Integrating Time (sec)*	*Minimum Signal Detectable (watt/m²)*	*Target*
Drake	NRAO (USA)	1960	13	1.42	100	350	100	10^{-21}	Epsilon Eridani, Tau Ceti
Troitsky, Rakhlin, Starodubtsev, Gershteir	Gorky Univ. (USSR)	1968	7.5	0.927	13	100	120	10^{-22}	12 nearby sunlike stars, Andromeda galaxy
Verschuur	NRAO (USA)	1972	21 45	1.42	3000 490	110	4000	10^{-26}	10 nearby sunlike stars
Troitsky, Bondar, Starodubtsev	Eurasian Network (USSR)	1970-present	dip.	1.875 1.00 0.60	700	600	150,000	?	Coincidental pulses from whole sky
Zuckerman, Palmer	NRAO (USA)	1972-1978	45	1.42	3000			10^{-26}	700 nearby sunlike stars
Bowyer, Zeitlin, Tarter, Lampton, Welch	Hat Creek (USA)	1972-present	13	var.	500	55	30	10^{-22}	Parasitic experiment on conventional research
Kardashev	Eurasian Network (USSR)	1972	dip.	1.337 – 1.863					Pulsed signals from whole sky
Bridle, Feldman	Algonquin (Canada)	1974-present	23	22.2	30,000			10^{-22}	70 nearby stars
Drake–Sagan	Arecibo (USA)	1975	150	1.42 1.653 2.38	1000			10^{-25}	Several nearby galaxies

Dixon, Cole	Ohio State Univ. (USA)	1973-present	transit telescope	1.42	20,000	100	10	10^{-21}	Entire sky
Wischnia	Copernicus satellite (USA)	1974		UV					Epsilon Eridani Tau Ceti Epsilon Indi
Cuzzi, Clark, Tarter, Black	NRAO (USA)	1976	21	8.522-8.523	5			2×10^{-24}	VLBI of 4 stars
Cuzzi, Clark, Tarter, Black	NRAO (USA)	1977	45	1.62-1.72	5.5	70	4-40	10^{-23}	200 sunlike stars
Drake, Stull	Arecibo (USA)	1977	150	1.42-1.66	10				Ten stars
Israel, Tarter	Westerbork Synthesis (Netherlands)	1975-79	12 × 25	1.42	$(4-10) \times 10^6$			7.7×10^{-22} to 6.4×10^{-24}	542 stars checked in continuous radio map
Shostak, Tarter	Westerbork Synthesis (Netherlands)	1975-79	12 × 25	1.42	$(4-10) \times 10^6$				Radio map of galactic center checked
Wielebinski, Sieradakis	Effelsburg (Germany)	1977-present	50	1.67					6 stars, pulses
Horowitz	Arecibo	1978	150	1.42	0.015	80	100	4×10^{-27}	200 sunlike stars
Cohen, Malkan, Dickey	Arecibo Massachusetts Australia	1978	150 18 31	1.67 22 1.61					25 globular star clusters
Billingham	Arecibo (USA)	1979	150	1.3-1.6	4.0		600		250 sunlike stars within 100 light-years of Sun

TABLE 5.2 (Continued)

Investigators	*Observatory*	*Date*	*Antenna Radius (m)*	*Freq. (GHz)*	*Band Width (Hz)*	*Det. Temp (K)*	*Integrating Time (sec)*	*Minimum Signal Detectable (watt/m²)*	*Target*
Cole, Ekers	Australia	1979							
Tarter, Duquet, Clark, Lesyna	Arecibo (USA)	1980-1981	150	1.42	5.5				210 sunlike stars
Biraud–Tarter	Nancay (France)	1980-		1.6					102 stars to date
Vallee, Simard–Normandin	Algonquin (Canada)	1982	23	10.5					Slow scan of sky for linearly polarized signals
Witteborn									20 stars
Lord, O'Dea	Mass.	1981	18	115					Search along galactic rotation axis at CO line
Horowitz, Teague, Linscott, Chen, Backus	Arecibo (USA)	1982	150		0.03				250 nearby stars
Horowitz	Oak Ridge (USA)	1983-?	13	var.	0.03				Scan of 68% of sky
Cyclops	?	?	20,000	1.42-1.66	0.1	20	2000	1×10^{-33}	All likely stars within 1000 light-years

Adapted from Dixon (1980); Morrison, Billingham, and Wolfe (1977); Murray, Gulkis, and Edelson (1978); Breuer (1982).

"glitches." This is the only way that it is possible to determine with any degree of certainty if an unusual radio signal is actually coming from outer space or simply a spurious response produced by terrestrial interference. The vast area of the array, plus the advanced electronics used for the receiving and data analysis system, should produce a detector which is a *billion* times more sensitive than the receiver used in Project Ozma. The cost of a full Cyclops array will probably exceed thirty billion dollars, which makes it as grandiose an undertaking as the Apollo moon-landing program.

FIGURE 5.6. Cyclops array of radio telescopes. From Oliver and Billingham (1973). National Aeronautics and Space Administration.

It will be necessary to build two Cyclops systems—one in the Northern Hemisphere and another in the Southern Hemisphere—in order to ensure full sky coverage. In order to spread out the cost as much as possible, the full system could be built up gradually, a few radio telescopes at a time, over a time period of ten to twenty years. The search could actually begin as soon as the first few radio telescopes were in place. The frequency band to be searched is the entire interval between 1.42 and 1.66 GHz—the "water hole." The computer system will be capable of resolving the incoming signal into 2.5 million individual frequency bands, each 0.1 Hz in width. This frequency resolution is near the maximum feasible for long-range interstellar communication. Approximately 2000 seconds (about a half-hour) will be spent listening to each candidate star, so about 15,000 stars could be examined in a year. At that rate, it could take several hundred years to search all of the likely stars within a thousand light-years of Earth. In order to avoid wasting time on stars that have little prospect of harboring intelligent civilizations, a parallel optical full-sky star survey is also proposed. This should turn up about 25 million prospects. The proposed Cyclops system would be so sensitive that it could detect a 25-megawatt omnidirectional beacon anywhere within a thousand light years of Earth. It could detect a beacon located on either Epsilon Eridani or Tau Ceti if it were as weak as 2500 watts! If there are any intelligent civilizations using microwave technology located within a thousand light-years of Earth, the Cyclops system should be able to spot them. With the addition of a few simple modifications, Cyclops could even be used as a highly powerful, long-range transmitter. If a total continuous power of a hundred megawatts is given to the entire array, a signal from the entire Cyclops system could be picked up at a range of 2000 light-years with only a primitive omnidirectional dipole antenna! It is so powerful that it could communicate with a duplicate facility located a billion light-years away in *another galaxy.*

The Sky's the Limit

NASA has also given some thought to an orbital SETI system, originally proposed by the Stanford Research Institute. It would have the advantage of operating in outer space and thereby relatively unhindered by terrestrial radio interference, an ever-present problem for ground-based SETI systems. Furthermore, since it would be located above the atmosphere it would be able to search in radio and microwave bands that are blocked by atmospheric absorption. A convenient location for the system might be at one of the stable Lagrange points along the Moon's orbital path. The proposed antenna would be a single reflective parabolic dish 3 kilometers across. It could be made of the flimsiest of materials, since it would not have to support any weight. The giant dish would be kept pointing away from Earth, so that no radio interference from satellites or other terrestrial transmitters could be inadvertently directed onto the detection system. As additional insurance, a large reflecting shield would be placed between the rear of the floating dish and the Earth. The dish would sweep 360 degrees of sky once every 28 days. Individual objects could be studied by moving a free-floating feed horn back and forth across the face of the dish, in much the same manner as the fixed Arecibo radio telescope operates. Any point in the universe could be examined at least once a month. The cost of an orbital SETI system is estimated to be about the same as that of the ground-based Cyclops system. The space shuttle could be used to lift components of the system into low-earth orbit, but some sort of "space tug" would have to be developed to move them out into near-lunar space where they would be assembled.

NASA has also given some consideration to another Stanford Research Institute idea, a SETI system located on the back side of the Moon. This system would have the additional advantage of being shielded from terrestrial radio interference by over 3000 kilometers of solid rock. The lunar system could either consist of a series of Arecibo-type fixed receivers built inside lunar craters, or it could actually involve a Cyclops-type orchard of fully steerable antennae. The cost of a lunar farside SETI system will of course be enormous, 200 billion dollars being probably a conservative estimate. Adding to the cost would be the sizeable lunar colony that would be required to support the system.

The Golden Fleece

In these times of tight budgets, such exotic and expensive SETI projects stand absolutely no chance of approval by governmental agencies. SETI advocates have been forced to be much more modest in their proposals. These have generally involved two different search philosophies. The first, originally pushed by the NASA-Ames Center at Moffett Field in California, is quite similar in approach to the Cyclops proposal, although on far less grandiose a scale. The Ames proposal involves a long-term radio search for signals from specific sunlike stars near Earth, using the water hole as the most promising frequency band. The 1000-foot Arecibo telescope, the 175-foot Ohio State University instrument, and a facility located in either Australia or Chile (to cover the Southern Hemisphere) would be used to make a narrow-bandwidth search of all 500 of the sunlike stars within 80 light years of Earth. The other approach was promoted by the Jet Propulsion Laboratory of Pasadena, California and is reminiscent of some of the early Soviet SETI efforts. It involves a search of the whole sky over relatively broad frequency bandwidths without attempting to preselect any particular targets. Several small antennae at the Goldstone spacetracking facility in California were to be used to search for intelligent signals over 100 MHz bandwidths within the frequency range between 1 and 25 GHz. A special multichannel analyzer was to be developed capable of chopping up the broad bandwidth into a million discrete channels, each 300 Hz in width. A hundred million sunlike stars could in principle be covered by the search, which was proposed to last as long as five years.

In 1977, NASA elected to follow a program which involved elements of both the Ames and JPL search philosophies. A long-term SETI program was finally approved with a yearly budget of 2.1

million dollars. The first five years of the project were to be taken up with equipment development, followed by a ten-year full-sky search over large frequency bandwidths using the small radio telescopes of the Goldstone Deep Space Network. A companion search, at greater sensitivity, of points of special promise was also planned, centered around the Arecibo radio telescope. The entire search program was to depend critically upon the development of a new type of multichannel spectrum analyzer, which in its final version would be able to observe 8 million frequency channels simultaneously with a broad range of choices of bandwidths. It was to be used first on the Goldstone telescopes and then moved to the Arecibo facility for the high-sensitivity phase of the search. Stanford University was given the responsibility for its development. Computer algorithms were to be designed to search the data for a variety of forms of intelligent radio signals. The program could eventually result in about a hundred million times more searching time than all previous SETI projects combined.

However, government-funded SETI efforts of any sort have recently encountered stiff opposition within certain sectors of the Congress. In 1978, Senator William Proxmire of Wisconsin gave NASA's SETI project a "Golden Fleece" award, which he regularly gives to government-funded research projects that he regards as of little value. As a result of the controversy the yearly budget of NASA's SETI program was cut back to $700,000, although enough additional funds were apparently transferred from other budgets to keep the project going. Senator Proxmire and his allies kept up the fire, one of their prime arguments being the "Fermi Paradox" which suggests that the current absence of extraterrestrials on Earth is a powerful argument for their complete nonexistence. If this argument is valid, any federally funded SETI project is obviously a waste of the taxpayer's money. Senator Proxmire finally had his way in 1981. In that year, any sort of new government funds specifically intended for SETI were officially eliminated from the budget, effectively killing the NASA program.

The intensity of the opposition to federally funded SETI efforts shocked many American scientists and astronomers. The author Carl Sagan, who has become somewhat of a publicist for the probable existence of intelligent life elsewhere in the universe, took up the crusade. He organized an extensive lobbying effort, circulated petitions, gathered signatures, and buttonholed legislators. He apparently succeeded in getting Senator Proxmire to mute his objections, and Congress decided to revive NASA's SETI project. The Congress freed up 1.5 million dollars for SETI to be spent in fiscal 1983. With this sort of funding available, NASA was able to revive its SETI effort.

An early prototype of Stanford's sophisticated multichannel signal processor (with 74,000 channels) was tested in October 1983. It is scheduled for testing on large radio telescopes in the near future. By the time that the 5-year development phase is over an 8-million channel analyzer will hopefully be available, capable of processing a billion bits of information per second. Signal processors capable of handling this enormous rate of information flow must also be developed. Armed with this advanced instrumentation, NASA will be able to begin its star search in 1988. In the first phase of the search, the instrumentation will be deployed on the small antennae of the Jet Propulsion Laboratory's Deep Space Network (located at sites in California, Spain, and Australia). These facilities will be used to sweep the entire sky over frequency bands ranging from 1 to 10 GHz, with some coverage between 10 and 25 GHz. This full-sky search will eventually explore virtually all of the microwave bands that are accessible from the ground and is scheduled to last at least three years. In the second phase of the search, the instrumentation will be attached to large, ultrasensitive radio telescopes (such as the Arecibo facility) to study individual stars. Present plans are to search 773 stars of classes F, G, and K within 60 light-years of the Sun over frequency bands between 1.2 to 3 GHz. In addition, a few spot bands between 3 and 25 GHz will be checked for unusual signals. By the time the program is completed, it will have been by far the most complete and thorough SETI effort ever carried out by humanity.

During the lean months in which no federal funding appeared to be forecoming for any sort of SETI activity, the irrepressible Carl Sagan joined with Bruce Murray (former head of the Jet Propulsion Laboratory) to seek other means by which even a modest interim search program could be carried out. They had founded a group known as the Planetary Society, a private philanthropic organization set up to provide some sort of limited financial support for solar system research and exploration in the face of the massive cuts being made in NASA's planetary probe budget. Paul Horowitz of Harvard University, together with some people from Stanford University and NASA-Ames, proposed that the 84-foot Oak Ridge, Massachusetts radio telescope operated by Harvard (which had not been used for several years) be reactivated and used for SETI. The Planetary Society agreed to fund this project to the tune of $20,000 per year. The radio telescope was provided with an advanced multichannel analyzer capable of resolving the incoming signal into 131,072 discrete bands, each 0.03 Hz wide. This receiving system was first tested at Arecibo in May 1982 during a search of 250 nearby sunlike stars, then moved to Massachusetts for permanent installation. Horowitz's SETI project was given the name "Project Sentinel." The first Sentinel search began in March of 1983 with appropriate fanfare and ceremony. Present plans call for a scan of 68 percent of the sky at a succession of different "magic frequencies," starting with the hydrogen line. No specific target stars will be studied, unless of course an "interesting" signal is found.

EAVESDROPPING ON THE UNIVERSE

The negative results of the early SETI efforts may mean that there are no technological civilizations in our immediate vicinity which are deliberately trying to establish radio contact with us. Nevertheless, there could be some civilizations which are not actively seeking to establish any sort of interstellar contact but are nevertheless "leaking" large amounts of radio energy into space in the natural course of their day-to-day activities. Do we have any chance of being able to detect such signals? Perhaps someone else in the stars has already picked *us* up. Earth has quite unintentionally been sending ever-increasing numbers of radio waves outward into space for much of the twentieth century, and it is conceivable that some alien civilization is at this very moment listening to our early radio broadcasts. Perhaps someone 40 to 50 light years away is now listening to the news broadcasts of the impending second world war and is deciding to tune us out, abandoning humanity as a doomed race.

Unintentional radio emissions from other civilizations will be far weaker than any signals deliberately sent to attract our attention. What are the prospects for our being able to detect such weak "leakage" signals? Since the only technological society we know of is our own, we can only answer this question by considering the prospects for the Earth's own radio emissions being picked up at interstellar distances. In 1978, a student-faculty team from the University of Washington undertook what was apparently the first systematic study of the radio "signature" of the Earth as viewed from interstellar space. The prospect of extraterrestrials currently listening to our early radio broadcasts is rather unlikely, since the entire AM radio band is strongly reflected by the ionosphere and never escapes into outer space except under highly unusual circumstances. Only those radio sources having frequencies above 20 MHz will be able to "leak" into outer space. The most powerful emitters at these frequencies are FM radio stations, television transmitters, various space surveillance systems operated by both the United States and the USSR, and several high-powered interplanetary radars used for astronomical research and space probe communication. Interplanetary radars, such as the one at Arecibo, are among the most powerful radio transmitters ever operated by humanity, but their operation is so intermittent in nature that their chance detection at interstellar distance would be extremely improbable. FM radio stations are in continual operation, but they are generally of such low power (by FCC decree) that any chance for detection at such large distances is rather poor. More promising sources are television transmitters and the American and Soviet versions of the Ballistic Missile Early Warning System (BMEWS).

A typical television station has a total power of about 100 kilowatts and operates in the frequency bands between 54 and 88 MHz (VHF, Channels 2 to 6), between 174 MHz and 216 MHz (VHF, Channels 7 to 13) or between 470 and 806 MHz (UHF, Channels 14 to 69). They send out signals having a total bandwidth of approximately 6 MHz. Such a wide bandwidth is required for the rapid transmission of the large amount of information that is contained in a single picture. However, about 90 percent of the signal's power is actually concentrated within a narrow "carrier" band only 0.1 Hz wide. In order to avoid wasting energy on signals needlessly sent into space, television transmissions are beamed toward a point on the local horizon of the station in a sheet about 5 degrees wide. This corresponds to a transmitter gain of about 25. A large fraction of this signal, however, misses the Earth altogether and leaks out into space. Could such a signal be picked up at interstellar distances? Because of its narrow bandwidth, the carrier component of a television signal has the best prospect of being detected. Surprisingly, the prospects for detection are rather poor. The galactic synchrotron radio noise effectively drowns out the lower-frequency VHF television stations (the noise temperature of the galaxy is 4500 kelvin at 40 MHz), leaving only the UHF stations. Calculations indicate that the Ozma receiver of 1960 could have detected a UHF TV station only if it were closer than 0.001 light-year to Earth. Even the highly sensitive SETI project of Horowitz, using the much larger Arecibo antenna, could have detected a UHF TV station at a maximum range of only a single light-year. This is not very encouraging, as the nearest stars are over four light-years away. In this light the negative results of the early SETI experiments should not be all that surprising, since even the Earth's television transmissions are undetectable at interstellar ranges using currently available equipment. In order to have any chances of picking up extraterrestrial television transmitters, Cyclops-sized arrays will be required. The proposed Cyclops system is so sensitive that it could spot a terrestrial UHF television transmitter at a range as great as 300 light-years. It is actually rather unlikely that even the Cyclops array could ever observe any sort of picture at such extreme distances, as that would require the use of much broader observational bandwidths and far shorter integrating times.

The prospects of intercepting a BMEWS transmission at interstellar distances may be more promising. The BMEWS system operates by sending out a powerful, fan-shaped radar beam into space. Any missile or satellite passing through the beam will reflect part of it back to receivers on the ground. The precise operational specifications of BMEWS radars are classified, but enough information has appeared in various trade journals so that some rough estimates of performance can be made. The operating frequencies of BMEWS radar installations are of the order of 400 MHz, and the gain of a typical transmitter is about a million. The peak beam power is of the order of 0.2 terawatt. The radar transmission consists of a sequence of intense pulses, each lasting about one millisecond with a bandwidth of approximately one kHz. These powerful radars are in continuous operation, but their frequencies are purposefully shifted in a random manner for security reasons. BMEWS radars are so powerful that the Ozma experiment could easily have detected such a transmitter at a range as large as 150 light-years. If such facilities had been operating on either Tau Ceti or Epsilon Eridani, Frank Drake should have been able to spot them. The proposed Cyclops system could detect a BMEWS installation located anywhere in the galaxy. Whether or not such a signal, once detected, would be recognized as artificial is another question. Regular sequences of radio pulses have been picked up by terrestrial receivers in the past, only upon closer examination to be ascribed to naturally occurring pulsars rather than to intelligent beings. Intense radio sources which vary rapidly in an irregular manner have been detected, but they are believed to have been produced by quasars rather than by Type III civilizations. It is possible that a signal from an extraterrestrial radar transmitter might someday be detected, only to be dismissed as just another pulsar, a new quasar, or other such naturally occurring object.

It is apparent that Cyclops-sized arrays will be required if there is to be any probability of picking up the leakage signals emitted by those extraterrestrial civilizations currently at a technological level roughly equivalent to present-day Earth. Perhaps more highly advanced civilizations which have reached the Type II or even the Type III level emit much larger amounts of "waste" radio energy, enough to be readily detectable by radio telescopes currently available on Earth. Carl Sagan has suggested that such prospects might actually be rather remote. The leakage of excessive amounts of radio energy into outer space is quite wasteful, and current trends on Earth seem to be away from long-range free-air broadcasting and toward cable or tight-beam microwave transmission of information. This is being done primarily to avoid overcrowding or saturation of the radio and microwave regions of the spectrum. Over the next few decades the amount of radio and microwave radiation which actually leaks out into space will probably undergo a steady decline, gradually reducing the radio signature of the Earth as seen from outer space. This will make it less and less likely in the future that we can be accidentally spotted by extraterrestrial radio astronomers. Furthermore, the costly and dangerous arms race must either voluntarily come to an end or else we will destroy ourselves. In either event, BMEWS radar installations will go off the air. Similar trends can be expected for other emerging Type I civilizations, leading to exceedingly short times during which they can be detected by spotting their accidental leakage signals. The average "leakage lifetime" could be as short as fifty to a hundred years, and as a result there will always be very few civilizations in this phase at any one time, even if the most favorable estimates for the numbers of habitable planets in the galaxy are accepted.

A MESSAGE FROM EARTH

In 1974, a series of major improvements was made to the facilities at the Arecibo radio telescope in Puerto Rico. Among these was the construction of equipment which made it possible to transmit a radio beam of unprecedented power. For short periods of time, a peak beam power of up to 20 terawatts could be produced. It was thought appropriate to celebrate the completion of these modifications by holding a unique ceremony. The radio telescope was used to send a short message to the edge of the galaxy, informing potential listeners of the existence of human life on Earth. The message was transmitted on November 16, 1974 at a frequency of 2380 MHz and a bandwidth of 10 Hz. The average beam power was about 3 terawatts, which made this particular signal perhaps the most powerful radio transmission yet made by the human race. The target was Messier 13, a globular cluster of approximately 300,000 stars in the constellation Hercules, 25,000 light-years distant from Earth. The apparent angular size of the cluster just about matches the angular dispersion of the beam, so that all of the stars in the cluster will be irradiated by the signal when it arrives.

The transmission of the information was done by successively switching the beam between two nearby frequencies so that a sequence of binary ones and zeroes was sent. A total of 1679 consecutive binary characters was transmitted, and the total duration of the message was 169 seconds. This left about a tenth of a second for each character, corresponding to the ten-Hertz bandwidth. The binary sequence that was actually transmitted is shown in Figure 5.7. It is actually a code for the construction of a two-dimensional picture. It is decoded by breaking up the message into 73 consecutive rows of 23

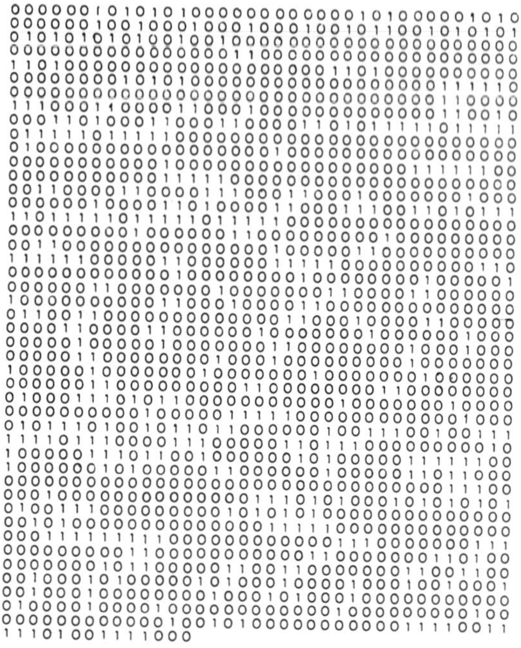

FIGURE 5.7. Binary array transmitted by Arecibo radio telescope in message sent to M13 in 1974. From Sagan and Drake (1975). Copyright © 1975, *Scientific American*, Inc. All rights reserved.

characters each and arranging each row one under the other. The zeroes represent blanks, whereas the ones represent darkened spaces. Seventy-three and twenty-three are prime numbers which might assist extraterrestrials in deciphering the message, but they probably could eventually arrive at the correct answer by simple trial and error. The message as properly laid out is shown in Figure 5.8.

The message starts with a counting lesson which describes the binary number system that is to be used. The numbers 1 through 10 are written across the top of the figure in binary notation, going from right to left. Each number has a "number label" attached to it at the bottom to show that it *is* a number as well as to indicate the direction from which it is to be read. The numbers 8, 9, and 10 are purposefully written on two different lines to indicate how numbers that are too large to appear on a single line are specified.

The next part of the message is an organic chemistry lesson. Just below the counting lesson is a diagram showing the five numbers 1,6,7,8, and 15. These are meant to stand for the atomic numbers of hydrogen, carbon, nitrogen, oxygen, and phosphorus, the prime ingredients of living matter on Earth. On the next few lines is a

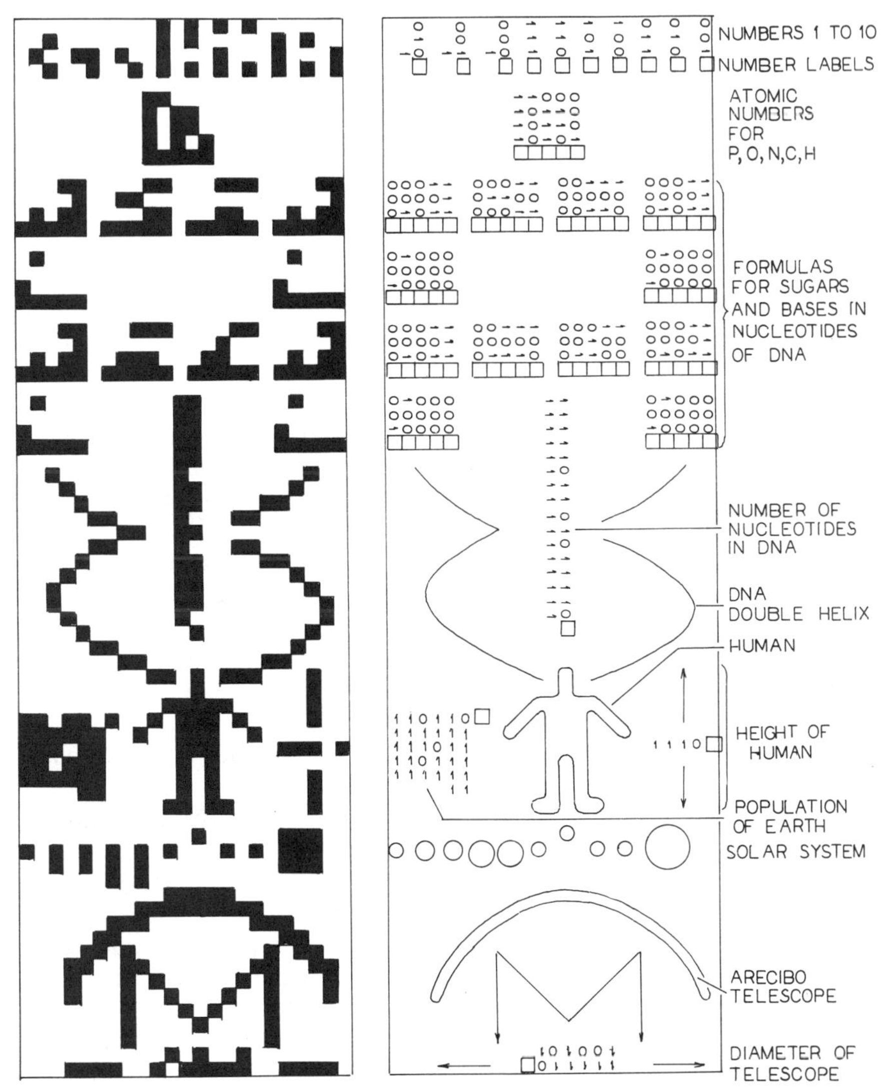

FIGURE 5.8. Layout and interpretation of Arecibo radio message to M13. From Sagan and Drake (1975). Copyright © 1975, *Scientific American,* Inc. All rights reserved.

numerical representation for the chemical formula of a chunk of double-stranded DNA. The right-hand number specifies the number of hydrogens, the next the number of carbons, etc., and each grouping of numbers is meant to represent a single molecule in the DNA chain. The actual physical structure of these molecules is not represented, but any extraterrestrials sufficiently advanced to be able to receive and decipher the message in the first place should also be able to deduce the molecular structures from a knowledge of the chemical formulas.

Below the chemical formula for DNA is a schematic view of this molecule when it is wound as a double helix. There is no number label on this particular figure, so it cannot be misread as some sort of number. In the middle of the helix there is a line of symbols with a number label attached. This number is 4 billion and is meant to represent the number of base pairs in the DNA of a single human chromosome. The double helix leads to the head of what is obviously a crude sketch of a human being. To the left of the human being is another large number (also about 4 billion), this time meant to denote the current earth population. To the right of the human is a height indicator with a small number (14) attached to it. The human is apparently 14 units tall. But what unit is used? The most logical unit of length is the wavelength of the microwaves used to send the message. This is 12.6 centimeters, so the human is 176 centimeters or 5 feet, 9 inches tall.

Underneath the human is a diagram of the solar system with the Sun on the right and the planets strung out to the left. The third planet is displaced upward toward the human, indicating that this planet is where the message came from. At the very bottom of the diagram is a crude sketch of the Arecibo radio telescope that transmitted the message. It is shown as being centered on the third planet, and the number across the bottom indicates that the radio telescope is 2430 wavelength units wide, or 1000 feet.

The transmission was so intense that radio receivers on planets within the Messier 13 system as small as 4 meters across could pick up the message when it reaches its destination 25,000 years hence. Alien radio astronomers observing the Sun at that particular time will see the radio luminosity of this otherwise quite ordinary star suddenly and inexplicably increase by seven orders of magnitude for a couple of minutes, momentarily outshining all other radio sources in the galaxy. The chance that any given radio telescope will just happen to be pointing in the right direction when the message arrives is of course almost zero, but the signal is probably powerful enough to be picked up by broad-area sky search equipment, such as the Eurasian Network of dipole antennas used by the Soviet Union in some of their SETI efforts. However, it is generally believed that there is very little chance that there is anyone actually living anywhere in the Messier 13 globular cluster who could pick up our signal when it does arrive. The stars within this cluster are Population II objects which are deficient in heavy metals and unlikely to have planets capable of sustaining any type of life at all, to say nothing of intelligent civilizations capable of interstellar communication.

THE GALACTIC CONNECTION

The first technological civilization in the galaxy may have appeared several billion years ago. By now, many millions of these societies must have come and gone in the long history of galactic life and civilization. Most probably perished very early in their careers, promptly using their newly developed technology to destroy themselves. They were unable to make any lasting imprint on the galaxy, leaving absolutely no trace of their existence before they vanished forever. A few civilizations, though, may have successfully overcome the pitfalls of technological adolescence to establish stable and enduring societies lasting for many thousands or even millions of years. These successful societies must have gained complete control over their home planets and perhaps even conquered their entire solar systems. Once this sort of mastery was achieved, they may have looked outward to the stars in search of new worlds to be occupied by their rapidly expanding societies. But there is absolutely no evidence for the current existence of any sort of galaxy-wide pattern of travel or trade that should have been the outcome of any truly successful attempt at interstellar space flight. In particular, there is no indication that any extraterrestrials have ever come to the Earth. In the words of Enrico Fermi: "Where are they?"

There are many alternative solutions to the Fermi Paradox that have been proposed, and we have discussed several of them in some detail. None is entirely satisfactory. We consider one of these possibilities in some detail here. This is the hypothesis that no civilization ever attempted large-scale interstellar travel because every one of them found it so much simpler and cheaper to interact with their neighbors via an exchange of radio signals rather than by making personal visits. Instead of an extensive pattern of interstellar travel, a vast network of civilizations in constant radio contact may have evolved. Extraterrestrials rarely if ever travel between the stars, but each communicative civilization in the galaxy may at this very moment be in contact with all the others by means of radio.

How could such a network of communicative civilizations ever become established? In the very early history of life in the galaxy there must have been only a very few communicative civilizations in existence, separated from each other by mean distances as large as several thousand light-years. How did these early societies ever manage to find one another? There are two distinct radio contact strategies that these pioneering technological societies might have chosen to follow, either passive or active. A passive search strategy would involve the use of sensitive receivers to listen for radio signals from alien intelligences, produced either by deliberate transmission or by unintentional "leakage." No transmission was attempted until the listeners had actually detected an intelligent signal from outer space. An active search strategy would involve the transmission of highly powerful "call signals" out into space in the hope of attracting the attention of anybody that might be listening.

The Long Search

Perhaps the first communicative civilization in the galaxy constructed a large array of radio telescopes like the proposed Cyclops system. The array was used to search all the sunlike stars within a range of a thousand light years for any radio "leakage" that might be evidence for the existence of a newly emerging technological society capable of radio contact. The construction and operation of a Cyclops-sized search system was a considerable investment of time and resources for any society, no matter how advanced. The initial cost of the system was no greater than the Apollo project which landed men on the Moon, but it had to be assumed that the search for alien technologies could last for many thousands or even millions of years. The first technological society to embark upon a systematic search for alien intelligence must have had a high degree of economic and political stability and virtually infinite patience as well to have operated such an expensive project for so long with absolutely no assurance of ultimate success.

The first searching civilizations may have listened to the stars for many millions of years before they found anything. Once success was achieved, the array of radio telescopes was used to direct a powerful microwave beam toward the newly discovered civilization, a signal so strong as to be readily detectable by even the most primitive electronic equipment. The first call signal probably carried an introductory message describing the nature of the society attempting contact and inviting the new society to respond. When the message finally arrived at its destination the target civilization suddenly noticed that it was being irradiated by an ultrastrong radio signal of obvious intelligent origin coming from outer space. After the message was decoded, a two-way conversation could begin. Once the first two technological societies were in contact, there would be additional impetus to find still more societies in the stars with which to converse. As new societies emerged one by one into the communicative phase, their leakage signals were spotted by the radio telescopes of those races already in contact. They were then invited to join the Network. Once the first few contacts were established, an expanding wave of communicating societies spread outward into the galaxy at nearly the speed of light. By now, the society of communicating worlds may have many millions of members. Perhaps our own radio leakage signals will soon impinge upon the receivers of an alien civilization, announcing our own readiness to join in the galactic dialogue.

The Starry Messenger

The galactic network could conceivably have been initiated by an active strategy. Perhaps the first technological civilization that appeared in the galaxy built an extremely powerful radio transmitter and used it to send an intense call signal into interstellar space in the hope that there was someone out there capable of answering. This beacon signal could have carried a message giving a brief description of the transmitting society and extending an invitation to respond. The first beacon transmitter must have operated for many millions of years before it was spotted by another intelligent society. Once the new society had detected the beacon, they could use conventional radio telescopes to answer. In this way, a two-way dialogue could begin. As each new society in the galaxy developed radio-astronomy technology, it noticed the beacon in the sky and joined in the cosmic dialogue. Perhaps it is only a matter of time before human radio astronomers spot the messenger in the stars that invites us to galactic citizenship.

One obvious choice for the beacon is a powerful, omnidirectional microwave transmitter strong enough to be spotted by anyone else in the galaxy having access to radio astronomy. Such an omnidirectional beacon has the obvious advantage that it beams its signal equally to all points in the sky, making certain that no one in the galaxy is unintentionally overlooked. However, there are significant drawbacks. For one, the amount of power needed for an omnidirectional transmitter to reach the most distant point in the galaxy with a detectable signal is enormous, as large as 10^{15} watts if operating at a bandwidth of 1 Hz. These sorts of power levels are three orders of magnitude greater than the total output of all the nations on Earth, and such energy expenditures would probably have to be borne for many millions of years before another civilization finally notices the transmissions. The costs involved in building and operating such a powerful facility are astronomical, and it may be true that no society, no matter how advanced, will long tolerate such high expenditures with absolutely no assurance of any ultimate benefit. Furthermore, these high power levels might actually be a health hazard for creatures living within the transmitting solar system. For these reasons such highly powerful, omnidirectional transmitters may be completely infeasible for even the most advanced technology.

Irrespective of their particular mode of operation, omnidirectional beacon transmitters are wasteful of energy, since they send out signals randomly into space, equally irradiating all points in the sky. Perhaps the first contacting civilization decided to be far more selective in its choice of targets. They could have built Arecibo-type radio telescopes and beamed powerful, highly directional signals toward target stars deemed to have a nonzero probability of having intelligent civilizations. By using directional transmitters, the call signals could easily be made strong enough to attract the attention of a civilization having access to even the most rudimentary radio technology. Powers of the orders of terawatts are nevertheless still required, and many millions of target stars must be addressed if there is to be any reasonable probability of contact. If each target star is irradiated for only a single hour, it could take over a hundred years to cover the million likely stars within a thousand light-years of the transmitting civilization, to say nothing of the rest of the galaxy. Each target star would have to be addressed many, many times if there is to be any chance of success.

An active search strategy involves some risks. For one, any sort of effective transmitter would by definition be a clear and obvious indicator of the exact location of the transmitting society. Interstellar transmitters could be an invitation to disaster at the hands of predatory alien civilizations waiting eagerly to pounce upon newly emerging technological societies. Even though interstellar imperialism is an extremely difficult undertaking, it is at least conceivable that there are some alien societies which might regard military conquest or economic exploitation of others as a tempting prospect. It may be unwise to advertise one's existence in such a blatant fashion without knowing what sorts of creatures are waiting out there. Second, a call signal sent out randomly into space may happen to fall upon an immature society which is intellectually or emotionally unprepared to face the fact that it is not alone in the universe. The sudden arrival of a powerful call signal of obvious intelligent origin might produce massive disruption. A premature revelation of the existence of alien intelligence could induce a catastrophic instability in the new society, creating so much confusion and disorder as to bring about its total collapse. For this reason, many societies may consider it immoral to broadcast signals randomly into space without knowing what sorts of disruptive effects they will have on other civilizations. As a result of these perceived dangers, no intelligent civilization may ever run the risk of operating long-range transmitters. We have never detected alien radio signals because no one is broadcasting. They are all waiting for a signal from *us*.

The Galactic Network

The odds against the establishment of a galactic network of communicative worlds seem to be enormous. It is truly the search for the proverbial needle in the haystack. The ultimate success of any search for extraterrestrial intelligence will depend upon the numbers of such societies currently inhabiting the galaxy, as well as on their mean separation. Frank Drake has pointed out that it would be impossible for a two-way communication network to exist if the mean distance between civilizations is so large that they perish before a radio signal has sufficient time to travel between them. Mathematically, the necessary criterion for the existence of a galactic communication network can be expressed as the requirement that the mean communicative lifetime (τ_{av}) be numerically at least twice as large as the time it takes for a light signal to travel between adjacent civilizations. This can be expressed as:

$$\tau_{av} \geq 2\ d_{av}/c \qquad (5\text{–}3)$$

where d_{av} is the mean separation between civilizations and c is the speed of light. By checking Table 3.1, we see that this condition is met only if the mean communicative lifetime τ_{av} is longer than several thousand years. Civilizations must be able to survive long enough for most to evolve into stable Type II interplanetary societies if there is to be any chance for the establishment of a two-way dialogue between worlds. Only if the probability for our own long-term survival is relatively high is there any chance for us being able to carry out conversations with others.

If there actually is a galactic network currently operating, can the Earth expect to join in? It is exceedingly unlikely that we might accidentally intercept a message sent between two worlds because such signals will be transmitted on beams of extremely narrow angular width to minimize the power required. The bandwidths of these signals will probably be quite broad in order to maximize the rate at which information is transmitted. No small portion of the spectra of these intelligent signals may appear to us to have any kind of order at all. Without knowledge of the coding structure, such a broadband intelligent signal may be indistinguishable from random background noise. If there are large numbers of advanced communicative civilizations in the galaxy, it is likely that they will detect our radio signals long before we detect theirs. The Earth has been sending out radio leakage signals into space at ever-increasing intensities for the past forty or fifty years, and every year that passes the chance increases that these signals will impinge upon the sensitive receivers of an extraterrestrial civilization. Perhaps someday quite soon a powerful radio signal will come to Earth from some point in the sky to tell us for the first time that we are not alone.

The Deprovincialization of Humanity

What will be the effect on human civilization when this signal arrives? How will humankind's view of its place in the cosmos be changed? After deciphering the signal, we may find that the extraterrestrials who sent the signal are so far advanced over us, both intellectually and technologically, that we are cast into a collective fit of despondency over our inferior state. We are the most junior member of the society of worlds. All of the scientific knowledge we had hoped to gain in the future is already well known to others. Most of the intellectual and social feats we had expected to perform in the future have already been accomplished by somebody else, in all probability in far more brilliant a fashion than we could ever hope to achieve. Mankind's thirst for increased knowledge and a better life could be severely dampened by the discovery of vastly superior societies in the stars; the essential human curiosity about the universe in which we live may be replaced by a passive inertia from which we may never recover. Conventional Western religions may be the first to suffer, as they have typically imagined mankind to be at the pinnacle of creation and the special recipient of God's attention. The knowledge that humanity is merely one of the more primitive intelligences in the universe may simply be too much for people with religious sensibilities to handle. The flying saucer and ancient astronaut cults will finally be thoroughly discredited, since there is obviously no reason for a civilization to send us radio messages when they are already dropping in personally on a regular basis. Dogmatic political and social philosophies may be especially hard hit. Imagine the effect on ultraconservatives in the United States if the extraterrestrial society turned out to be largely socialist or collectivist in orientation. Or imagine the effect on Communists in easter Europe or the Soviet Union if the aliens stressed individual initiative and free enterprise. However, it is far more likely than not that extraterrestrial civilizations have political and social structures so utterly different from anything that has ever appeared on Earth that the terms we use to describe human patterns of organization simply have no meaning within the context of an extraterrestrial society. Extraterrestrials may have no concept of territoriality, property, money, religion, government, status hierarchies, bureaucracy, economic competition, or a whole host of other ideas and institutions that are so much a part of human society. We may actually find some of their practices to be offensive; they could conceivably engage in cannibalism, infanticide, ritual murder, or other such habits which we find loathsome. Their personal appearance may be repulsive to us; they might actually physically resemble such obnoxious terrestrial creatures as cockroaches or slugs. Their sexual mores may be entirely different from our own; indeed, they may not even have sexes at all, reproducing via means more bizarre than we can imagine. Their concept of good and evil may not be the same as our own; they may not even have any concept of good or evil at all. The notion of good versus evil may be a strictly human idea and not universal to all intelligent creatures. We must be prepared to reexamine all of our preconceived notions about what is "proper" behavior for intelligent beings and what is the "correct" mode of organization for civilization. At the very least, interstellar contact should be culturally enlightening.

Bearing in mind all the differences certain to exist between their society and ours, do we nonetheless respond to an invitation to talk? Some scientists have seriously suggested that we dare not answer any radio message lest it be a lure sent out by predatory aliens to trick us into exposing our world to conquest. Advanced alien civilizations may deem our planet a convenient target for economic exploitation; they may be running out of energy and material resources on their own world and could be tempted to seize the resources on ours. They could conceivably wish to enslave humans, to use us as pets or experimental animals, perhaps even to harvest us as a sort of gourmet delicacy. They may have a chauvinistic need to expand their life and culture throughout the galaxy at the expense of others by the use of subversion and trickery. Are such fears justified? In 1971, a group of Soviet and American scientists interested in the question of extraterrestrial life met in a conference at Byurakan in Soviet Armenia. Among other things, they considered the response that humanity should make to a deliberate message from an extraterrestrial society. Because of the potential dangers involved, the participants all agreed that the entire international community at the highest level must be consulted before any decision is made about a response. No nation, group, or individual should act unilaterally. The Arecibo message of 1974, as well as the Pioneer and Voyager records and plaques, were apparently violations of this accord, as no prior consultation was made before they were sent into interstellar space. This caused some consternation among a few of the conference participants, but as these messages were not responses to an extraterrestrial signal there was not too much concern expressed. However, we should dampen our enthusiasm for extraterrestrial contact with a little caution in the future.

If we do decide to respond, what do we tell them? Presumably, we could give a brief description of our solar system, the nature of our home planet, biochemical information about our type of life, details about our general physical appearance, perhaps even some indication of the nature of our human society and our psychological makeup. In formulating our reply, it will be necessary to think of mankind in the cosmic sense as we attempt to give some sort of objective description of ourselves that would be understood by an alien intelligence. We should be frank about our weaknesses as well as our accomplishments; we remain divided among ourselves and are still struggling against the dangers posed by our own technological and emotional immaturity. Perhaps it might help if we sent them the plays of Shakespeare as well as scientific or astronomical information, the music of Bach as well as biological data. But we must bear in mind that such things as poetry, drama, music, and art may have significance only to us human beings and may make no sense to an extraterrestrial. It is especially important that we make it clear to the aliens that we are interested only in a relatively limited dialogue with them and that we insist on retaining our distinctly human identity, even though they may be far advanced over us both intellectually and technologically. We do not wish to be colonized or exploited by anyone, nor do we wish our culture to be subordinated to an overpowering galactic authority.

What results could we anticipate from a long-term dialogue with extraterrestrials? We should not expect any instant answers to our pressing scientific, sociological, political, or economic problems. The nearest alien society is likely to be sufficiently distant that any answer to our queries will be hundreds or thousands of years in coming. By the time that the answer finally arrives we are likely to have either forgotten the question or else to have figured it out for ourselves many years earlier. Two hundred years ago the most important questions in the physical sciences were the phlogiston theory of matter and the caloric theory of heat. The scientific fads of today (quarks and black holes) may sound just as foolish two hundred years from now. The scientific knowledge of the aliens is likely to be incomprehensible if presented to us in bulk, just as a twentieth-century textbook on quantum mechanics would have been incomprehensible to even the best minds of ancient Greece. Our form of life is likely to be sufficiently different from theirs that their advanced biological knowledge would be of little aid to us in curing diseases or in prolonging life. Aliens know so little of human psychology that they can tell us nothing that would be of any significant help in preventing us from destroying ourselves. Our particular problems are ours and ours alone to solve; the aliens can give us some historical perspectives from other societies that were faced with similar problems and perhaps some advice and suggestions, but the bulk of the task is up to us. We can expect no free gifts from the stars.

Even though we should not expect too much from a dialogue with extraterrestrials, entrance into the galactic network could be of inestimable benefit to mankind. It might be a shock to our egos to find that there are vastly superior beings in the stars, but we humans are sufficiently resilient creatures so that we would probably quickly get over our initial feelings of inferiority. Human culture is likely to be sufficiently different from theirs to have particularly unique features not reproduced anywhere else in the galaxy. We will be valuable and worthwhile members of the galatic community for that reason alone. Even though humanity has a lot to learn from extraterrestrials, there is much that we can teach them as well. Instead of a massive inferiority complex, interstellar contact might produce a resurgence of pride in simply being human and a putting aside of the petty squabbles that currently divide mankind into warring camps. The mere existence of long-lived extraterrestrial societies would prove that it is indeed possible to overcome the perils that currently threaten our existence, provided we exert our full effort and ingenuity to the task at hand. If others have survived, perhaps we too have a chance.

Perhaps large numbers of extraterrestrial civilizations are currently involved in long-term radio conversations, but no one is particularly interested in talking to *us.* Alien societies may be carrying out a busy conversation with their peers, but they are so far superior to contemporary human society that they are no more interested in a dialogue with Earth than we would be interested in an intellectual discourse with a single-celled protozoan. They may be perfectly aware of the existence of humanity, but they regard us as unworthy of radio contact. There may be a "threshold" of intellectual and technological development which we must exceed before anyone else in the stars will pay any particular attention to us. We may have to wait many years before we are deemed worthy of membership in the society of communicative worlds.

Man, the Galactic Pioneer

We may wait in vain for that first message from the stars that invites us to galactic citizenship. If the mean communicative lifetime is as short as a century, we could be the only communicative society currently inhabiting the galaxy. Many other societies have preceded us, but all perished very early in the Type I phase before they had much of a chance to seek out other civilizations around other stars.

Alternatively, the galaxy may currently be in the very earliest era of the history of life, and the human race is among the first intelligent species to have appeared. Humanity is then faced with a unique opportunity; we may be the society destined to play the key role in establishing the galactic network.

A model of the evolution of galactic intelligence proposed by Sebastian von Hoerner of NRAO suggests that the initial establishment of two-way interstellar contact is the crucial catalyst that permits the eventual emergence of technological societies having long lives. Throughout all the previous history of the galaxy every society has lived separately, without any form of assistance from its neighbors. Unaided, they were all at the mercies of their own weaknesses and inadequacies and destroyed themselves soon after they attained Type I status. However, if two or more societies can establish radio contact they can provide each other with valuable advice and assistance in overcoming the problems that they share in common, vastly improving the odds for the long-term survival of them all. The mean communicative lifetime of galactic civilizations may undergo a radical increase at the instant of first radio contact. Our very survival on Earth may depend upon our finding someone else in the stars to talk to. A search for extraterrestrial intelligence may be the most important and significant action in which the human race could conceivably be involved.

Being the initiator of interstellar contact is comforting and reassuring, since we would have the maximum freedom of action. This will carry with it an equal moral responsibility. There has been some discussion of the moral and ethical considerations involved in humanity's first contact with extraterrestrials. This has generally been under the aegis of what has come to be known as "metalaw," the hypothetical body of moral principles supposedly applicable to all intelligent creatures. Many terrestrial cultures have been decimated by ill-advised attempts by technologically superior societies to "civilize" them, and we do not want such disasters to be repeated on a cosmic scale. Robert A. Freitas insists that we must never attempt to contact another culture that might not survive the shock of finding out that we exist. We must carefully observe before we attempt any contact. Call signals should then never be sent out randomly into space; the only ethical means of contact would be a passive radio search in which a new society is carefully studied before any overtures are made. The Austrian "space lawyer" Ernst Fasan warns that when the contact attempt is actually made we must be careful not to demand impossible or harmful acts of the new civilization, nor must we appear to be threatening their living space or the uniqueness of their culture. This again requires careful precontact observation, as extraterrestrials are certain to have completely different psychological drives than humans. Barbara Moskowitz argues that we must always respect the free choice of extraterrestrial societies. Any society must have the right to accept or reject any advances that are made toward it. If, after hearing our call signal, they decide not to have anything to do with us, we must respect their wishes, ill-advised though we think them to be. Andrew Haley proposes that the "Golden Rule" of "do unto others as you would have them do unto you" may not be a valid guide for dealing with extraterrestrials, as they are likely to have entirely different needs and desires than humans. He suggests that a better guide might be the principle of "do unto others as they would have you do unto them."

After many years of listening we may find that the search for extraterrestrial intelligences is in vain. If we are alone, that too is a fact worth knowing. The Earth is then a unique spot in the universe; only here on this planet have intelligent creatures evolved, capable of looking out into space and wondering where they fit into the overall plan of the cosmos. We are more precious than any of the classical regions have dared imagine, and there is an even greater obligation placed upon the human race to survive and ensure that the first emergence of consciousness into the universe is not snuffed out in its infancy.

APPENDIXES

APPENDIX 1

Geological Time Intervals

Cryptozoic Eon (3800-700 million years ago)
  Hadean Era (4450 million-3800 million years ago)
  Archean Era (3800 million-2500 million years ago)
  Proterozoic Era (2500 million-700 million years ago)
Phanerozoic Eon (700 million years ago-present)
  Paleozoic Era (700-230 million years ago)
    Eocambrian Period (700-560 million years ago)
    Cambrian Period (560-485 million years ago)
    Ordovician Period (485-435 million years ago)
    Silurian Period (435-410 million years ago)
    Devonian Period (410-350 million years ago)
    Mississippian Period (350-320 million years ago)
    Pennsylvanian Period (320-290 million years ago)
    Permian Period (290-230 million years ago)
  Mesozoic Era (230-65 million years ago)
    Triassic Period (230-192 million years ago)
    Jurassic Period (192-135 million years ago)
    Cretaceous Period (135-65 million years ago)
  Cenozoic Era (65 million years ago-present)
    Paleogene Period (65-25 million years ago)
      Paleocene Epoch (65-54 million years ago)
      Eocene Epoch (54-36 million years ago)
      Oligocene Epoch (36-25 million years ago)
    Neogene Period (25 million years ago-present)
      Miocene Epoch (25-6 million years ago)
      Pliocene Epoch (6-2 million years ago)
      Pleistocene Epoch (2 million years ago-10,000 years ago)
      Holocene Epoch (10,000 years ago-present)

APPENDIX 2

The Time Scale of Evolution
(1G – 1 billion years, 1M = 1 million years, 1K = 1 thousand years)

Era	Time	Event
CRYPTOZOIC	5G	Protosolar nebula formed
		Formation of earth
		Interior heating
		Formation of core and mantle
		First atmosphere and oceans
		Oldest rocks yet found
	4G	End of meteorite impact era
		Loss of highly reducing atmosphere
		Origin of life
		Anaerobic photosynthesis
	3G	First permanent continental crust. First supercontinent.
		Nitrogen-fixing bacteria.
		Cyanobacteria-O_2 level begins to rise
	2G	Eucaryotic cells
	1G	Sexual reproduction
		Multicellular life.
	600M	Gondwanaland formed
		First nervous systems
		Invertebrate animals with external skeletons
PALEOZOIC	500M	Jawless fish
		N. America collides with Europe
		Ordovician Ice Epoch
		Ray-finned fish
		Lower plants on land
		Lungfish
	400M	Insects and higher plants on land
		Large coastal forests
		Jawed fishes
		Appalachian mountains formed
		First amphibians
		First seed-bearing plants
		Diversification of amphibians
		Diversification of insects
		Vast swampy forests
	300M	First reptiles
		Gondwanan Ice Epoch
		Diversification of reptiles
		Reptiles dominant on land
		Pangaea formed
		Large-scale extinction
		N. America separates from Africa
		First dinosaurs
MESOZOIC	200M	First mammals
		Large amphibians die out
		First birds
		Conifers dominant plants on land
		Dinosaurs dominant on land
		First flowering plants
		Diversification of insects
	100M	India separates from Antarctica
		S. America separates from Africa
	70M	

Epoch	Date	Event
PALEOCENE		Late Cretaceous extinction
	60M	Mammals begin to diversify
		Early Rocky Mountains
		N. America separates from Europe
		First horses
EOCENE	50M	Australia quits Antarctica
		India collides with Asia
	40M	S. America leaves Antarctica
		First monkeys, first cats
OLIGOCENE	30M	Early pigs and bears
		Whales appear; first grasses
		First apes
	20M	Evolution of ground-feeding apes
		Swiss Alps form
MIOCENE	10M	
	6M	Separation of ape and hominid lines
	5M	Himalayan mountains formed
PLIOCENE	4M	Modern horses
		Australopithecus afarensis
	3M	N. and S. America rejoin
		Australopithecus africanus
	2M	Australopithecus robustus/boisei
		Homo Hablis
		Quaternary Ice Epoch begins
		First stone tools and hunting
		Homo Erectus

Epoch	Date	Event
PLEISTOCENE	1M	Illinoisian Ice Age
	100K	Neanderthal Man
	80K	Wurm Ice Age
	60K	
	40K	Cro-Magnon Man
		First permanent shelters
		Last ice age begins
		First boats
		First men in Australia and in Americas
HOLOCENE	20K	Height of ice age
		Bow and arrow
		Domestication of animals
		Ice age comes to end
	10K	Development of agriculture
		First towns
		First metallurgy
		First cities
	8K	First pottery
		Effective copper metallurgy
	6K	Urbanization, taxes, class divisions
		Invention of wheel
		First pictographs
		First writing
		First civilizations in Near East
		Pyramids
		First civilization in India
		First empires
		Chariots
	4K	First civilizations in China
		Iron smelting
		Mayan civilization in America
		Golden Age of Greece
		Origin of science and

	philosophy
	Water power
2K	Roman Empire established
	Roman Empire collapses
1000	Rise of Islamic science
	Managed economies
	Magnetic compass
800	Rise of Aztecs
	Gun warfare
600	Moveable-type printing press
	Discovery of New World
	Circumnavigation of globe
	Evolution of modern nation-state in Europe
400	Invention of telescope
	Imperial nationalism
	Political liberalism
	First steam engine
	Newtonian mechanics
	Discovery of oxygen
	Beginning of industrialism
200	Revolutionary nationalism
	First steamships and railroads
	Germ theory of disease
	Maxwell's Equations
	Theory of Evolution
100	Internal combustion engine
	Telephone, electric light
	Automobile
	Discovery of radioactivity
	Invention of radio
80	First airplane
	Special theory of relativity
	Submarine, gas, tank, air warfare
	First communist governments
60	First commercial radio broadcasts
	Quantum theory
	First liquid-fuelled rocket
	First commercial TV broadcasts
	First jet aircraft
	Global total warfare
40	V-2 rocket; electronic computer
	Atomic bomb
	Radar signal bounced off Moon
	Transistor
	Hydrogen bomb; gene structure
	Radar signal bounced off Venus
	First earth satellite
	First lunar probes; Laser
	First man in space
20	First planetary probes
	Communication satellites
	Manned landing on Moon
	First probe to leave solar system
	First radio signal beamed out of solar system
	Gene-splicing
	Reusable spaceship
0	

APPENDIX 3 The Moons of the Solar System

Name	Semi-Major Axis (km)	Eccentricity	Orbital Inclination wrt Equator of Primary (deg)	Sid. Rot. Period (days)	Sid. Orb. Period (days)	Radius (km)	Mass (Moon = 1)	Density (gm/cc)
Earth								
Moon	384,400	0.055	var	27.32	27.32	1738	1.00	3.34
Mars								
Phobos	9378	0.015	1.02	0.319	0.319	9.0x10.5x13.5	1.0×10^{-6}	2.0
Deimos	23,459	0.00005	1.82	1.262	1.262	7.5x6.0x5.0	1.3×10^{-7}	1.9
Jupiter								
Metis (J16)	128,240	?	?	0.295	0.295	20	?	?
Adrastea (J14)	128,930	?	?	0.297	0.297	12	?	?
Amalthea (J5)	161,140	0.003	0.46	0.418	0.418	75x85x135	?	?
Thebe (J15)	222,330	?	1.25	0.675	0.675	35-40	?	?
Io (J1)	421,600	0.000	0.03	1.769	1.769	1815	1.21	3.53
Europa (J2)	670,900	0.000	0.47	3.551	3.551	1570	0.66	3.03
Ganymede (J3)	1,070,000	0.001	0.18	7.155	7.155	2630	2.03	1.93
Callisto (J4)	1,883,000	0.007	0.25	16.689	16.689	2400	1.45	1.79
Leda (J13)	11,110,000	0.147	27	?	240	3-8	?	?
Himalia (J6)	11,476,000	0.158	28	0.5(?)	250.6	50-60	?	?
Lysithea (J10)	11,700,000	0.12	29	?	259.2	7-8	?	?
Elara (J7)	11,737,000	0.207	26	?	259.7	12-20	?	?
Anake (J12)	21,200,000	0.169	147	?	630(R)	6-8	?	?
Carme (J11)	22,600,000	0.207	163	?	692(R)	7-10	?	?
Pasiphae (J8)	23,500,000	0.40	147	?	739(R)	6-10	?	?
Sinope (J9)	23,600,000	0.275	156	?	758(R)	7-10	?	?
Saturn								
Atlas (S17)	137,670	0.002	0.3	0.602	0.602	20x10x?	?	?
1980S27 (S16)	139,400	0.003	0.0	0.613	0.613	70x50x40	?	?

1980S26 (S15)	141,700	0.004	0.05	0.629	0.629	55x45x35	?	?
Epimetheus (S11)	151,422	0.009	0.34	0.694	0.694	70x60x50	?	?
Janus (S10)	151,472	0.007	0.14	0.695	0.695	110x100x80	?	?
Mimas (S1)	188,224	0.0201	1.52	0.964	0.964	195	0.00051	1.2
1980S12 (?)	188,200	?	?	?	0.964	5	?	?
Enceladus (S2)	240,192	0.0044	0.023	1.389	1.389	250	0.0010	1.1
Tethys (S3)	296,563	0.000	1.09	1.906	1.906	525	0.0083	1.0
1980S34 (?)	?	?	?	?	?	5	?	?
Telesto (S13)	296,560	?	?	?	1.906	17x14x13	?	?
Calypso (S14)	296,560	?	?	?	1.906	17x11x11	?	?
1981S6 (?)	296,560	?	?	?	1.906	?	?	?
1981S10 (?)	350,000	?	?	?	2.44	5	?	?
1980S6 (S12)	378,600	0.005	0.15	?	2.739	18x16x15	?	?
1981S7 (?)	378,600	?	?	?	?	5	?	?
Dione (S4)	379,074	0.0022	0.0023	2.756	2.756	560	0.0140	1.4
1981S9 (?)	470,000	?	?	?	3.8	5	?	?
Rhea (S5)	527,828	0.0010	0.35	4.528	4.528	765	0.033	1.3
Titan (S6)	1,221,432	0.0290	0.33	?	15.938	2570	1.85	1.92
Hyperion (S7)	1,502,275	0.1042	0.28-0.93	?	21.739	115x145x190	?	?
Iapetus (S8)	3,559,400	0.0283	15(v)	79.243	79.243	720	0.026	1.2
Phoebe (S9)	12,900,000	0.1633	150	0.39	550.4(R)	110	?	?
Uranus								
Miranda	130,000	0.000	0	1.414(R)	1.414(R)	120-200	?	?
Ariel	192,000	0.003	0	2.520(R)	2.520(R)	600-730	?	?
Umbriel	267,000	0.004	0	4.144(R)	4.144(R)	505-605	?	?
Titania	438,000	0.002	0	8.706(R)	8.706(R)	740-860	?	?
Oberon	586,000	0.001	0	13.463(R)	13.463(R)	745-885	?	?
Neptune								
Triton	355,000	0.0(?)	160	5.876(R)	5.876(R)	1600-2640	2.0(?)	?
Nereid	5,562,000	0.75	28	?	359.88	140-300	0.0007(?)	?
Pluto								
Charon	17,000	?	?	6.4(?)	6.4	675-900	0.02(?)	?

R = Retrograde. v = Variable. Mass of Earth's Moon = 7.354×10^{22} kilograms. Adapted from Mitton (1977).

APPENDIX 4

Table of Planetary Properties

	Mercury	*Venus*	*Earth*	*Mars*	*Jupiter*	*Saturn*	*Uranus*	*Neptune*	*Pluto*
Semimajor axis (AU)	0.39	0.72	1.00	1.52	5.20	9.5	19.2	30.1	39.72
Eccentricity	0.206	0.007	0.017	0.093	0.049	0.054	0.047	0.009	0.25
Inclination wrt Ecliptic (degrees)	7.0	3.39	0.00	1.85	1.30	2.49	0.77	1.77	17.2
Siderial period (years)	0.241	0.615	1.00	1.88	11.86	29.46	84.01	164.79	250.3
Siderial rotation period (days)	58.65	243.01	0.997	1.026	0.415*	0.445*	0.62 – 1.0 (R)	0.75-1.0	6.4
Inclination of equator to Orbital Plane (degrees)	7.0	177.4	23.5	23.98	3.08	26.73	97.92	28.8	50(?)
Equatorial radius (km)	2440	6070	6378	3389	71,540	58,200	25,700-27,900	25,000(?)	1500-1800
Ellipticity	0	0	0.003	0.009	0.064	0.10	0.02-0.03	0.02(?)	?
Mass (Earth masses)	0.056	0.82	1.00	0.108	318	95.1	14.5	17.2	0.002
Average density (gm/cc)	5.43	5.25	5.50	3.93	1.33	0.71	1.24	1.67	0.4-1.0
Effective temperature (C)	430	−29	−20	−33	−149	−179	−215	−218	−220 (d)
Average surface temp. (C)	430 (d) −170 (n)	460	15-20	−33 (d) −85 (n)	no surface	no surface	no surface	no surface	−229 (d) −270 (n)
Surface pressure (atm)	zero	90	1.0	0.008	no surface	no surface	no surface	no surface	0.0001-0.05
Magnetic moment (gauss-cm^3)	2.4-6.0 × 10^{22}	$<10^{22}$	7.91 × 10^{25}	<2.5 × 10^{21}	1.55 × 10^{30}	4.7 × 10^{28}	?	?	?
Atmospheric constituents (by weight percent)	none	97 CO_2, 3 N_2, H_2O, Ar, CO	78 N_2, 21 O_2, 1 Ar, CO_2, H_2O	95 CO_2, 2.5 N_2, 1.5 Ar	81 H_2, 19 He	89 H_2, 11 He	89 H_2, 11 He	? H_2, ? He, ? CH_4	CH_4

R = retrograde.
* = radio period.

mass of earth = 5.98 × 10^{24} kilograms.
1 AU = 1.497 × 10^{11} meters.

APPENDIX 5

Properties of the Sun

Mass	1.991×10^{30} kilograms
Mean earth-sun distance	1.49×10^{11} meters = 1 astronomical unit (AU)
Radius	6.960×10^{8} meters
Density	1.41 grams/cc (average) 150 grams/cc (core)
Rotational period	24 days 16 hours (at equator)
Rotational velocity	2.06 kilometers per second (at equator)
Escape velocity from surface	617 kilometers per second
Gravitational acceleration at surface	273.7 meters per second^2 (27.9 g)
Angle of inclination between equator and ecliptic	7 degrees
Oblateness	None observable
Solar luminosity (L_o)	3.9×10^{26} watts
Average power emitted per unit surface area	6.44×10^{7} watts per square meter
Solar constant (sunlight intensity at Earth's orbit)	1.353×10^{3} watts per square meter (± two percent)
Effective surface temperature	5500 degrees Celsius
Composition (atomic percent)	93.9 percent H 5.9 percent He rest heavier elements

APPENDIX 6

The Galaxy

Stars:	2 × 10^{11}, of which about 50 percent are double or multiple. One third in Population 1, two thirds in Population II
Mass:	1.4 × 10^{12} solar masses (1 × 10^{6} solar masses in nucleus)
Absolute magnitude:	−20.5
Total luminosity:	4 × 10^{37} watts
Radius:	80,000 light-years (3 light-years for nucleus) Sun 28,000 light-years from center Corona perhaps out as far as 300,000 light-years
Thickness:	3000 light-years (stellar disk) 300 light-years (gaseous disk) 6000 light-years (bulge at galactic center)
Galactic Arms:	1500 light-years wide, 600 light-years thick Contain many regions of dust and many new star clusters
Rotation:	One rotation every 50,000 years (galactic nucleus) One rotation every 225 million years (Sun's orbit)
Rate of new star creation:	15 per year, on the average.
Gas:	10^{6}-10^{8} atoms per cubic meter inside gaseous disk 73 percent hydrogen, 25 percent helium, rest heavier elements.

Adapted from Mitton (1977) and Icke (1975).

APPENDIX 7 Atomic Abundances

Element	Atomic Number	Percentage Abundance by Number of Atoms					
		Universe	*Sun*	*Entire Earth*	*Earth's Crust*	*Ocean Water*	*Human Body*
Hydrogen	1	92.7	93.9	0.12	2.88	66.2	60.56
Helium	2	7.2	5.9	—	—	—	—
Lithium	3	—	—	—	0.009	—	—
Beryllium	4	—	—	—	—	—	—
Boron	5	—	—	—	—	—	—
Carbon	6	0.008	0.039	0.099	0.055	0.0014	10.68
Nitrogen	7	0.015	0.008	0.0003	0.007	—	2.44
Oxygen	8	0.050	0.065	48.88	60.43	33.1	25.67
Fluorine	9	—	—	0.0038	0.077	—	—
Neon	10	0.020	0.003	—	—	—	—
Sodium	11	0.0001	0.0002	0.64	2.55	0.29	0.075
Magnesium	12	0.0021	0.004	12.5	1.78	0.034	0.011
Aluminum	13	0.0002	0.0003	1.3	6.25	—	—
Silicon	14	0.0023	0.0042	14.0	20.48	—	—
Phosphorus	15	—	—	0.14	0.079	—	0.13
Sulfur	16	0.0009	0.0015	1.4	0.033	0.017	0.13
Chlorine	17	—	—	0.045	0.011	0.34	0.033
Argon	18	0.0003	0.0001	—	—	—	—
Potassium	19	—	—	0.056	1.374	0.006	0.037
Calcium	20	0.0001	0.0002	0.46	1.878	0.006	0.23
Scandium	21	—	—	—	—	—	—
Titanium	22	—	—	0.028	0.19	—	—
Vanadium	23	—	—	—	0.004	—	—
Chromium	24	—	—	—	0.008	—	—
Manganese	25	—	—	0.056	0.037	—	—
Iron	26	0.0014	0.003	18.87	1.858	—	—

A blank indicates that the abundance is less than 0.001 percent. Adapted from Dickerson (1978) and Pasachoff (1973).

APPENDIX 8 A List of Molecules Found Within the Interstellar Medium

Formula	*Name*	*Date Discovered*	*Formula*	*Name*	*Date Discovered*
CH	methylidyne	1940(opt)	C_2H	ethynyl	1974
CN	cyanogen	1940(opt)	N_2H	protonated nitrogen	1974
CH^+	methylidyne ion	1941(opt)	CH_3NH_2	methyl amine	1974
OH	hydroxyl	1963	CH_3CH_3O	dimethyl ether	1974
NH_3	ammonia	1968	NS	nitrogen sulfide	1975
H_2CO	formaldehyde	1969	SO_2	sulfur dioxide	1975
H_2O	water	1969	NH_2CN	cyanimide	1975
H_2	hydrogen	1970(uv)	CH_2CHCN	vinyl cyanide	1975
CO	carbon monoxide	1970	$HCOOCH_3$	methyl formate	1975
HCO^+	formyl ion	1970	C_2H_5OH	ethanol	1975
CH_3OH	methanol	1970	HCO	formyl	1976
CS	carbon monosulfide	1971	C_2H_2	acetylene	1976
SiO	silicon monoxide	1971	HC_5N	cyanodiacetylene	1976

HCN	hydrogen cyanide	1971	C_2	diatomic carbon	1977(ir)
HNC	hydrogen isocyanide	1971	HNO	nitroxyl	1977
OCS	carbonyl sulfide	1971	C_3N	cyanoethynyl	1977
HCOOH	formic acid	1971	CH_2CO	ketene	1977
HC_3N	cyanoacetylene	1971	CH_4	methane	1978
CH_3CN	methyl cyanide	1971	C_4H	butadiynyl	1978
NH_2CHO	formamide	1971	CH_3CH_2CN	ethyl cyanide (proprionitrile)	1978
H_2S	hydrogen sulfide	1972			
HNCO	isocyanic acid	1972	HC_7N	cyanotriacetylene	1978
SO	sulfur monoxide	1973	HC_9N	cyano-octatetrayne	1978
H_2CS	thioformaldehyde	1973	HNCS	isothiocyanic acid	1979
CH_2NH	methanimine	1973	CH_3SH	methyl mercaptan	1979
CH_3C_2H	methyl acetylene	1973	O_3	ozone	1980
CH_3CHO	acetaldehyde	1973	$HC_{11}N$	cyanopentaacetylene	

opt = first discovered by optical spectrum
uv = first discovered by ultraviolet spectrum
ir = first discovered by infrared spectrum

All other molecules listed were discovered by microwave emission or absorption.

Based on data from A. P. C. Mann and D. A. Williams, *Nature* (London) *283,* 721 (1980).

REFERENCES

ALVAREZ, L. W., W. ALVAREZ, F. ASARO, and H. V. MICHEL, "Extraterrestrial Cause for the Cretaceous-Tertiary Extinction," *Science,* 208 (1980), 1095–108.

ARRHENIUS, S., "The Propagation of Life in Space," *Die Umschau,* 7 (1903), 481–82.

ARVIDSON, R. E., A. B. BINDER, and K. L. JONES, "The Surface of Mars," *Scientific American,* 238, No. 3 (1978), 76–89.

ASCHER, R., and M. ASCHER, "Interstellar Communication and Human Evolution," *Nature,* 193 (1962), 940–41.

ASIMOV, I., *Asimov's Guide to Science.* New York, N. Y.: Basic Books, Inc., 1972.

———, *Extraterrestrial Civilizations.* New York, N. Y.: Crown, 1979.

BAKER, D., "Report From Jupiter," *Spaceflight,* 16 (1974), 140–44; 17 (1975), 102–7.

BALL, J. A., "The Zoo Hypothesis," *Icarus,* 19 (1973), 347–49.

———, "Extraterrestrial Intelligence: Where Is Everybody?," *American Scientist,* 68, (1980), 656.

BAROSS, J. A., and J. W. DEMING, "Growth of "Black Smoker" Bacteria At Temperatures at Least 250C," *Nature,* 303 (1983), 423–24.

BARTLETT, A. A., "Forgotten Fundamentals of the Energy Crisis," *American Journal of Physics,* 46, no. 9 (1978), 876–88.

BATES, D. R., "Difficulty of Interstellar Radio Communication," *Nature,* 248 (1974), 317–18.

———, "CETI: Put Not Your Trust in Beacons," *Nature,* 252 (1974), 432–33.

BEATTY, J. K., "NASA's SETI: Bigger is Better," *Technology Review,* 87, no. 1 (1984), 56–57.

BEATTY, J. K., B. O'LEARY, and A. CHAIKIN, *The New Solar System.* Cambridge, Mass.: Sky Publishing Co., 1981.

BEEBE, R., "Planetary Atmospheres," *Reviews of Geophysics and Space Physics,* 21 (1983), 143–51.

BIERI, R., "Humanoids on Other Planets?," *American Scientist,* 52 (1964), 452–58.

BLACK, D. C., and G. C. J. SUFFOLK, "Concerning the Planetary System of Barnard's Star," *Icarus*, 19 (1973), 353–57.

BLITZ, L., M. FICH, and S. KULKARNI, "The New Milky Way," *Science*, 220 (1983), 1233–40.

BLUM, H. F., "Dimensions and Probability of Life," *Nature*, 206 (1965), 131–32.

BOK, B. J., "The Milky Way Galaxy," *Scientific American*, 244, no. 3 (1981), 92–120.

BOND, A., and A. R. MARTIN, "Project Daedalus: The Mission Profile," *Journal of the British Interplanetary Society*, 29 (1976), 101–12.

———, "A Conservative Estimate of the Number of Habitable Planets in the Galaxy," *Journal of the British Interplanetary Society*, 31 (1978), 411–15.

BONDY, S. C., and M. E. HARRINGTON, "L Amino Acids and D Glucose Bind Stereospecifically to a Colloidal Clay," *Science*, 203 (1979), 1243–44.

BOWYER, S., and others, "The Berkeley Parasitic SETI Program," *Icarus*, 53 (1983), 147–55.

BRACEWELL, R. N., "Communications from Superior Galactic Communities," *Nature*, 186 (1960), 670–71.

———, *The Galactic Club: Intelligent Life in Outer Space*. San Francisco, Ca.: San Francisco Book Co., 1976.

BRACHET, J., "The Living Cell," *Scientific American*, 205, no. 3 (1961), 51–61.

BREUER, R., *Contact with the Stars: The Search for Extraterrestrial Life*. San Francisco, Ca.: W. H. Freeman, 1982.

BROWN, R. L., K. J. JOHNSTON, and K. Y. LO, "High-Resolution VLA Observations of the Galactic Center," *The Astrophysical Journal*, 250 (1981), 155–59.

BUSSARD, R. W., "Galactic Matter and Interstellar Flight," *Astronautica Acta*, 6, no. 4 (1960), 179–94.

CALVIN, M., "Chemical Evolution," in *Intestellar Communication*, ed. A. G. W. Cameron. New York, N. Y.: Benjamin, 1963.

———, "Chemical Evolution," *American Scientist*, 63 (1975), 169–77.

CAMERON, A. G. W., Ed., *Interstellar Communication*. New York, N. Y.: Benjamin, 1963.

———, "Communicating With Intelligent Life on Other Worlds," *Sky and Telescope*, 26 (1963), 258–61.

CARR, M. H., "The Volcanoes of Mars," *Scientific American*, 234, no. 1 (1976), 33–43.

———, "The Geology of the Terrestrial Planets," *Reviews of Geophysics and Space Physics*, 21, no. 2 (1983), 160–72.

CARVER, J. H., "Prebiotic Atmosphere Oxygen Levels," *Nature*, 292 (1981), 136–38.

CASSEN, P., R. T. REYNOLDS, and S. J. PEALE, "Is There Liquid Water on Europa?," *Geophysical Research Letters,* 6, no. 9 (1979), 731–34.

CASSENTI, B. N., "A Comparison of Interstellar Propulsion Methods," *Journal of the British Interplanetary Society,* 35 (1982), 116–24.

CHAPMAN, C. R., *Planets of Rock and Ice: From Mercury to the Moons of Saturn.* New York, N. Y.: Charles Scribner's Sons, 1982.

CHAPMAN, R. D., *Discovering Astronomy.* San Francisco, Ca.: W. H. Freeman, 1978.

CHAPPELL, W. R., R. R. MEGLEN, and D. D. RUNNELLS, "Comments on Directed Panspermia," *Icarus,* 21 (1974), 513–15.

CLARKE, J. N., "Extraterrestrial Intelligence and Galactic Nuclear Activity," *Icarus,* 46 (1981), 94–96.

CLOUD, P., "The Biosphere," *Scientific American,* 249, no. 3 (1983), 176–89.

COCCONI, G., and P. MORRISON, "Searching for Interstellar Communication," *Nature,* 184 (1959), 844–46.

COLE, G. H. A., "The Internal Structure and Early History of the Moon," *Journal of the British Astronomical Association,* 90, no. 6 (1980), 539–59.

———, "Aspects of the Physics of Planetary Interiors," *Contemporary Physics,* 22, no. 4 (1981), 397–424.

COX, L. J., "An Explanation for the Absence of Extraterrestrials on Earth," *Quarterly Journal of the Royal Astronomical Society,* 17, no. 2 (1976), 201–8.

CRICK, F. H. C., and L. E. ORGEL, "Directed Panspermia," *Icarus,* 19 (1973), 341–46.

CROWLEY, T. J., "The Geologic Record of Climatic Change," *Reviews of Geophysics and Space Physics,* 21, no. 4 (1983), 828–77.

DAEDALUS STUDY GROUP, "Project Daedalus," *Spaceflight,* 19 (1977), 419–30.

DE SAN, M. G., "The Ultimate Destiny of an Intelligent Species—Everlasting Nomadic Life in the Galaxy," *Journal of the British Interplanetary Society,* 34 (1981), 219–37.

DICKERSON, R. E., "Chemical Evolution and the Origin of Life," *Scientific American,* 239, no. 3 (1978), 70–86.

DIXON, R. S., "A Search Strategy for Finding Extraterrestrial Radio Beacons," *Icarus,* 20 (1973), 187–99.

———, "The Sentinel," *Cosmic Search,* 2, no. 1 (1980), 40–45.

DIXON, R. S., and D. M. COLE, "A Modest All-Sky Search for Narrow-band Radio Radiation Near the 21-cm Hydrogen Line," *Icarus,* 30 (1977), 267–73.

DOLE, S. H., *Habitable Planets For Man.* New York, N. Y.: Blaisdell, 1964.

DONAHOE, F. J., "On the Abundance of Earth-Like Planets," *Icarus*, 5 (1966), 303–4.

DONAHUE, T. M., and others, "Venus Was Wet: A Measurement of the Ratio of Deuterium to Hydrogen," *Science*, 216 (1982), 630–33.

DOTT, R. H., and R. L. BATTEN, *Evolution of the Earth*. New York, N. Y.: McGraw-Hill, 1971.

DRAKE, F. D., "How Can We Detect Radio Transmission From Distant Planetary Systems?" *Sky and Telescope*, 19 (1960), 140–43.

———, "Project Ozma," *Physics Today*, 14, no. 2 (1961), 40–46.

———, "On the Abilities and Limitations of Witnesses of UFOs and Similar Phenomena," in *UFOs—A Scientific Debate*. eds. C. Sagan and T. Page. New York, N. Y.: W. W. Norton, 1972.

———, "A Reminiscence of Project Ozma," *Cosmic Search*, 1, no. 1 (1979), 10–15.

DYSON, F. J., "Search for Artificial Stellar Sources of Infrared Radiation," *Science*, 131 (1960), 1667–68.

———, "Interstellar Transport," *Physics Today*, 21, no. 10 (1968), 41–45.

EDELSON, E., *Who Goes There? The Search for Intelligent Life in the Universe*. New York, N. Y.: Doubleday, 1979.

EDELSON, R. E., "An Observational Program to Search for Radio Signals from Extraterrestrial Intelligence Through the Use of Existing Facilities," *Acta Astronautica*, 6 (1979), 129–43.

EDWARDS, I. E. S., *The Pyramids of Egypt*. New York, N. Y.: Viking, 1972.

EHRLICH, P., and others, "Long-Term Biological Consequences of Nuclear War," *Science*, 222 (1983), 1293–1300.

ENGEL, M. H., and B. NAGY, "Distribution and Enantiomeric Composition Of Amino Acids in the Murchison Meteorite," *Nature*, 296 (1982), 837–40.

ESHLEMAN, V. R., G. F. LINDAI, and G. L. TYLER, "Is Titan Wet or Dry?" *Science*, 221 (1983), 53–55.

FINK, U., and others, "Detection of a CH_4 Atmosphere on Pluto," *Icarus*, 44 (1980), 62–71.

FITCH, F. W., and E. ANDERS, "Observations on the Nature of the 'Organized Elements' in Carbonaceous Chondrites," *Annals of the New York Academy of Sciences*, 108 (1963), 495–513.

FLASAR, F. M., "Oceans on Titan?" *Science*, 221 (1983), 55–57.

FLOWER, A. R., "World Oil Production," *Scientific American*, 238, no. 3 (1978), 42–49.

FORWARD, R. L., "A Programme for Interstellar Exploration," *Journal of the British Interplanetary Society*, 29 (1976), 611–32.

———, "Antimatter Propulsion," *Journal of the British Interplanetary Society*, 35 (1982), 391–95.

FOWLER, J. M., *Energy and the Environment.* New York, N. Y.: McGraw-Hill, 1975.

FREEMAN, J., and M. LAMPTON, "Interstellar Archaeology and the Prevalence of Intelligence," *Icarus,* 25 (1975), 368–69.

FREITAS, R. A., "Interstellar Probes: A New Approach to SETI," *Journal of the British Interplanetary Society,* 33 (1980), 95–100.

———, "A Self-Reproducing Interstellar Probe," *Journal of the British Interplanetary Society,* 33 (1980), 251–64.

FREITAS, R. A., and F. VALDES, "A Search for Natural or Artificial Objects Located at the Earth-Moon Libration Points," *Icarus,* 42 (1980), 442–47.

FROUDE, D. O., and others, "Ion Microprobe Identificaton of 4100-4200 Myr-Old Terrestrial Zircons," *Nature,* 304 (1983), 616–18.

GARDNER, M., *Fads and Fallacies in the Name of Science,* New York, N. Y.: Dover, 1957.

GARRISON, W. M., and others, "Reduction of Carbon Dioxide in Aqueous Solutions by Ionizing Radiation," *Science,* 114, (1951), 416–18.

GATEWOOD, G., "On the Astrometric Detection of Neighboring Planetary Systems," *Icarus,* 27 (1976), 1–12.

GEBALLE, T. R., "The Central Parsec of the Galaxy," *Scientific American,* 241, no. 1 (1978), 60–70.

GLIESE W., and H. JAHREISS, "Nearby Star Data Published 1969–1978," *Astronomy and Astrophysics Supplement,* 38 (1979), 423–48.

GOLDSMITH, D., and T. OWEN, *The Search for Life in the Universe.* Menlo Park, Ca.: Benjamin/Cummings, 1980.

GRIFFITH, E. E., and A. W. CLARKE, "World Coal Production," *Scientific American,* 240, no. 1 (1979), 38–47.

GRIFFITH, J. S., "A Guide to Stellar Objects," *Spaceflight,* 12 (1970), 213–19.

GUALTIERI, D. M., "Trace Elements and the Panspermia Hypothesis," *Icarus,* 30 (1977), 234–38.

HALDANE, J. B. S., "Origin of Life," in *The Rationalist Annual.* London, England: The Rationalist Press Association, 1929.

HALLAM, A., "Alfred Wegener and the Hypothesis of Continental Drift," *Scientific American,* 232, no. 2 (1975), 88–97.

HARRINGTON, R. S., and B. J. HARRINGTON, "Can We Find a Place to Live Near a Multiple Star?" *Mercury,* 7, no. 2 (1978), 34–37.

HART, M. H., "An Explanation for the Absence of Extraterrestrials on Earth," *Quarterly Journal of the Royal Astronomical Society,* 16, no. 2 (1975), 128–35.

———, The Evolution of the Atmosphere of the Earth," *Icarus,* 33 (1978), 23–39.

———, "Habitable Zones About Main Sequence Stars," *Icarus,* 37 (1979), 351–57.

———, "Was the Pre-Biotic Atmosphere of the Earth Heavily Reducing?" *Origins of Life,* 9 (1979), 261–66.

———, "Atmospheric Evolution, the Drake Equation, and DNA: Sparse Life in an Infinite Universe," in *Extraterrestrials—Where are They?* eds. M. H. Hart and B. Zuckerman. Elmsford, N. Y.: Pergamon Press, 1982.

HART, M. H., and B. ZUCKERMAN, Eds. *Extraterrestrials—Where are They?.* Elmsford, N. Y.: Pergamon Press, 1982.

HARTMANN, W. W., *Astronomy: The Cosmic Journey.* Belmont, Ca.: Wadsworth, 1978.

HARVEY, P. M., B. A. WILKING, and M. JOY, "On the Far-Infrared Excess of Vega," *Nature,* 307 (1984), 441–42.

HEAD, J. W., C. A. WOOD, and T. A. MUTCH, "Geologic Evolution of the Terrestrial Planets," *American Scientist,* 65 (1977), 21–29.

HEAD, J. W., S. E. YUTER, and S. C. SOLOMON, "Topography of Venus and Earth: A Test for the Presence of Plate Tectonics," *American Scientist,* 69 (1981), 614–23.

HENDERSON-SELLERS, A., and J. G. COGLEY, "The Earth's Early Hydrosphere," *Nature,* 298, (1982), 832–35.

HEPPENHEIMER, T. A., *Toward Distant Suns: A Bold, New Prospectus for Human Living in Space.* Harrisburg, Pa.: Stackpole Books, 1979.

HODGE, P. W., *Concepts of Contemporary Astronomy.* New York, N. Y.: McGraw-Hill, 1974.

HOHLFIELD, R. G., and Y. TERZIAN, "Multiple Stars and the Number of Habitable Planets in the Galaxy," *Icarus,* 30 (1977), 598–600.

HOROWITZ, N. H., "The Search for Life in the Solar System," *Accounts of Chemical Research,* 9, no. 1 (1976), 1–7.

———, "The Search for Life on Mars," *Scientific American,* 237, no. 5 (1977), 52–61.

HOROWITZ, P., "A Search for Ultra-Narrowband Signals of Extraterrestrial Origin," *Science,* 201 (1978), 773–35.

HOYLE, F., and C. WICKRAMASINGHE, *Lifecloud: The Origin of Life in the Universe.* New York, N. Y.: Harper and Row, 1978.

HUANG, S., "Occurrence of Life in the Universe," *American Scientist,* 47 (1959), 397–402.

———, "The Sizes of Habitable Planets," *Publication of the Astronomical Society of the Pacific,* 72 (1960), 489–94.

———, "Extrasolar Planetary Systems," *Icarus,* 18 (1973), 339–76.

HUBBARD, W. B., "Interiors of the Giant Planets," *Science,* 214 (1981), 145–49.

———, "Constraints on the Origin and Interior Structure of the Major Planets," *Philosophical Transactions of the Royal Society of London,* A303 (1981), 315–26.

HUBBERT, M. K., "The Energy Resources of the Earth," *Scientific American,* 225, no. 3 (1971), 60–70.

HUGHES, D. W., "Comets," *Contemporary Physics,* 23, no. 3 (1982), 257–83.

HUNTEN, D. M., "The Outer Planets," *Scientific American,* 223, no. 3 (1975), 131–40.

ICKE, V., "The Exploration of Our Galaxy," *Spaceflight,* 17 (1975), 414–20.

INGERSOLL, A. P., "Jupiter and Saturn," *Scientific American,* 245, no. 6 (1981), 90–108.

———, "The Atmosphere," *Scientific American,* 249, no. 3 (1983), 162–74.

IRVINE, W. M., S. B. LESCHINE, and F. P. SCHLOERB, "Thermal History, Chemical Composition and Relation of Comets to the Origin of Life," *Nature,* 283 (1980), 748–49.

ISBELL, W. H., "The Prehistoric Ground Drawings of Peru," *Scientific American,* 239, no. 4 (1978), 140–53.

JASTROW, R., *Until the Sun Dies.* New York, N. Y.: W. W. Norton, 1977.

———, *Red Giants and White Dwarfs.* New York, N. Y.: W. W. Norton, 1979.

———, *The Enchanted Loom: Mind in the Universe.* New York, N. Y.: Simon and Schuster, 1981.

JOHNSON, T. V., and L. A. SODERBLOM, "Io," *Scientific American,* 249, no. 6 (1983), 56–67.

JONES, D. M., "A New Possibility for CETI," *Spaceflight,* 19 (1977), 113–14.

JONES, E. M., "Discrete Calculations of Interstellar Migration and Transport," *Icarus,* 46 (1981), 328–36.

KARDASHEV, N. S., "Transmission of Information by Extraterrestrial Civilizations," *Soviet Astronomy-AJ,* 8, no. 2 (1964), 217–21. Translated from *Astronomicheskii Zhurnal,* 41, no. 2 (1964), 282–87.

KASTING, J. F., S. C. LIU, and T. M. DONAHUE, "Oxygen Levels in the Prebiological Atmosphere," *Journal of Geophysical Research,* 84, no. C6 (1979), 3097–107.

KAUFMAN, W. J., *Astronomy: The Structure of the Universe.* New York, N. Y.: Macmillan, 1977.

KLEIN, H. P., "The Viking Mission and the Search for Life on Mars," *Reviews of Geophysics and Space Physics,* 17, no. 7 (1979), 1655–62.

KOPAL, Z., *The Realm of the Terrestrial Planets.* New York, N. Y.: Wiley, 1979.

KRAUS, J., "We Wait and Wonder," *Cosmic Search,* 1, no. 3 (1979), 31–34.

KREIFELDT, J. G., "A Formulation for the Number of Communicative Civilizations in the Galaxy," *Icarus,* 14 (1971), 419–30.

KUIPER, T. B. H., "Galactic-Scale Civilization," in *Strategies for the Search for Life in the Universe,* ed. M. D. Papagiannis. Dordrecht, Holland: D. Reidel Publishing Co., 1980.

KUIPER, T. B. H., and M. MORRIS, "Searching for Extraterrestrial Civilizations," *Science,* 196 (1977), 616–21.

LAWTON, A. T., "The Nearest Other Solar System?," *Spaceflight,* 12 (1970), 170–73.

———, Interstellar Communication—Antenna or Artifact?" *Journal of the British Interplanetary Society,* 27 (1974), 286–94.

LEHNINGER, A. L., *Biochemistry.* New York, N. Y.: Worth, 1970.

LEOVY, C. B., "The Atmosphere of Mars," *Scientific American,* 237, no. 1 (1977), 34–43.

LEVINE, J. S., "The Photochemistry of the Paleoatmosphere," *Journal of Molecular Evolution,* 18 (1982), 161–72.

LEVINE, J. S., and F. ALLARIO, "The Global Troposphere: Biogeochemical Cycles, Chemistry, and Remote Sensing," *Environmental Monitoring and Assessment,* 1 (1982), 263–306.

LIPPINCOTT, S. L., "Astrometric Search for Unseen Stellar and Sub-Stellar Companions to Nearby Stars and the Possibility of Their Detection," *Space Science Reviews,* 22 (1978), 153–89.

LOEFFLER, F., "Is Anybody There?" *Nature,* 303 (1983), 467.

MALLOVE, E. F., "Renaissance in the Search for Galactic Civilizations," *Technology Review,* 87, no. 1 (1984), 48–54.

MANN, A. P. C., and D. A. WILLIAMS, "A List of Interstellar Molecules," *Nature,* 283 (1980), 721–25.

MARION, J. B., *Energy in Perspective.* New York, N. Y.: Academic Press, 1974.

MARKOWITZ, W., "The Physics and Metaphysics of Unidentified Flying Objects," *Science,* 157 (1967), 1274–79.

MARTIN, A. R., "The Detection of Extrasolar Planetary Systems," *Journal of the British Interplanetary Society,* 27 (1974), 881–906.

MARTIN, A. R., and A. BOND, "Nuclear Pulse Propulsion: A Historical Review of an Advanced Propulsion Concept," *Journal of the British Interplanetary Society,* 32 (1974), 283–310.

———, "Is Mankind Unique?—The Lack of Evidence for Extraterrestrial Intelligence," *Journal of the British Interplanetary Society,* 36 (1983), 223–25.

MARX, G., "Interstellar Vehicle Propelled by Terrestrial Laser Beam," *Nature,* 211 (1966), 22–23.

MATTINSON, H. R., "Project Daedalus: Astronomical Data on Nearby Stellar Systems," *Journal of the British Interplanetary Society,* 29 (1974), 76–93.

MAYR, E., "Evolution," *Scientific American*, 239, no. 3 (1978), 47–55.

MAZUR, P., and others, "Biological Implications of the Viking Mission to Mars," *Space Science Reviews*, 22 (1978), 3–34.

MEINSCHEIN, W., B. NAGY, and D. J. HENNESSY, "Evidence in Meteorites of Former Life: The Organic Compounds in Carbonaceous Chondrites are Similar to Those Found in Marine Sediments," *Annals of the New York Academy of Sciences*, 108 (1963), 553–79.

MENDELSSOHN, K., "A Scientist Looks at the Pyramids," *American Scientist*, 59 (1971), 210–20.

MENZEL, D. H., "UFOs—The Modern Myth," in *UFOs—A Scientific Debate*, eds. C. Sagan and T. Page. New York, N. Y.: W. W. Norton, 1972.

MICHAUD, M. A. G., "Two Tracks to New Worlds," *Spaceflight*, 19 (1977), 2–6.

MILLER, R., and W. K. HARTMANN, *The Grand Tour: A Traveler's Guide to the Solar System*. New York, N. Y.: Workman, 1981.

MILLER, S. L., "The Formation of Organic Compounds on the Primitive Earth," *Annals of the New York Academy of Sciences*, 69 (1957), 260–75.

MILLER, S. L., and H. C. UREY, "Organic Compound Synthesis on the Primitive Earth," *Science*, 130 (1959), 245–51.

MITTON, S., *The Cambridge Encyclopaedia of Astronomy*. New York, N. Y.: Crown, 1977.

MOLTON, P. M., "On the Likelihood of a Human Interstellar Civilization," *Journal of the British Interplanetary Society*, 31 (1978), 203–8.

MOORE, P., and G. HUNT, *Atlas of the Solar System*. Chicago, Ill.: Rand McNally and Co., 1983.

MORRISON, P., "Interstellar Communication," *Bulletin of the Philosophical Society of Washington*, 16 (1962), 58–81.

MURRAY, B. C., "Mercury," *Scientific American*, 233, no. 3 (1975), 59–68.

MURRAY, B., S. GULKIS, and R. E. EDELSON, "Extraterrestrial Intelligence: An Observational Approach," *Science*, 199 (1978), 485–92.

NEWMAN, M. J., and R. T. ROOD, "Implications of Solar Evolution for the Earth's Early Atmosphere," *Science*, 198 (1977), 1035–37.

OLIVER, B. M., "State of the Art in the Detection of Intelligent Extraterrestrial Signals," *Astronautica Acta*, 18 (1973), 431–39.

———, "Proximity of Galactic Civilizations," *Icarus*, 25 (1975), 360–67.

OLIVER, B. M., and J. BILLINGHAM, codirectors: "Project Cyclops: A Study of A System for Detecting Extraterrestrial Life," NASA CR 114445.

O'Neill, G. K., "The Colonization of Space," *Physics Today,* 27, no. 9 (1974), 32–40.

Oriti, R. A., and W. B. Starbird, *Introduction to Astronomy.* Encino, Ca.: Glencoe Press, 1977.

Oro, J., "Studies in Experimental Organic Cosmochemistry," *Annals of the New York Academy of Sciences,* 108 (1963), 464–81.

Oro, J., and A. P. Kimball, "Synthesis of Purines Under Possible Primitive Earth Conditions," *Archives of Biochemistry and Biophysics,* 94 (1961), 217–27.

Owen, T., "Titan," *Scientific American,* 246, no. 2 (1982), 98–109.

———, "Life as a Planetary Phenomenon," *Earth-Oriented Applications of Space Technology,* 4 (1984), 31–38.

Papagiannis, M. D., "Could We Be the Only Advanced Technological Civilization in the Galaxy?" in *Proceedings of the Second ISSOL Meeting,* ed. H. Noda. Tokyo, Japan: Center for Academic Publications, 1978.

———, "Liquid Water on a Planet Over Cosmic Periods," in *Origin of Life,* ed. Y. Wolman. Dordrecht, Holland: D. Reidel, 1981.

———, "The Search for Extraterrestrial Civilizations—A New Approach," *Mercury,* 11, no. 1 (1982), 12–25.

———, "The Colonization of The Galaxy—A Key Concept in the Search for Extraterrestrial Intelligence," in *Compendium in Astronomy,* ed. E. G. Mariolopoulos and others. Dordrecht, Holland: D. Reidel, 1982.

———, Ed., *Strategies for the Search for Life in The Universe.* Dordrecht, Holland: D. Reidel, 1980.

Parker, E. N., "The Sun," *Scientific American,* 233, no. 3 (1975), 43–50.

Pasachoff, J. M., *Astronomy Now.* Philadelphia, Pa.: W. B. Saunders, 1973.

Pettengill, G. H., D. B. Campbell, and H. Masursky, "The Surface of Venus," *Scientific American,* 243, no. 2 (1980), 54–65.

Phillips, R. J., and others, "Tectonics and Evolution of Venus," *Science,* 212 (1981), 879–87.

Pollack, J. B., "Mars," *Scientific American,* 233, no. 3 (1975), 107–17.

Ponnamperuma, C., and A. G. W. Cameron, *Interstellar Communication: Scientific Perspectives.* Boston, Mass.: Houghton Mifflin, 1974.

Ponnamperuma, C., and P. Kirk, "Synthesis of Deoxyadenosine Under Simulated Primitive Earth Conditions," *Nature,* 203 (1964), 400–401.

Ponnamperuma, C., and others, "Formation of Adenine by Electron Irradiation of Methane, Ammonia, and Water," *Proceedings of the National Academy of Sciences,* 49 (1963), 737–40.

POLLARD, W. G., "The Prevalence of Earthlike Planets," *American Scientist,* 67 (1979), 653–59.

RAO, U. R., "Geomagnetic Field—Its Role in the Evolution of Life and Intelligence on Earth," *Journal of the British Interplanetary Society,* 34 (1981), 459–65.

RAYMO, Chet, *Biography of a Planet.* Englewood Cliffs, N. J.: Prentice-Hall, Inc., 1984.

———, *The Crust of Our Earth.* Englewood Cliffs, N. J.: Prentice-Hall, Inc., 1983.

———, *365 Starry Nights.* Englewood Cliffs, N. J.: Prentice-Hall, Inc., 1982.

REYNOLDS, R. T., and P. M. CASSEN, "On the Internal Structure of the Major Satellites of the Outer Planets," *Geophysical Research Letters,* 6, no. 2 (1979), 121–24.

RIDPATH, I., *Worlds Beyond: A Report on the Search for Life in Space.* New York, N. Y.: Harper and Row, 1978.

———, *Messages from the Stars: Communication And Contact With Extraterrestrial Life.* New York, N. Y.: Harper and Row, 1978.

ROOD, R. S., and J. S. TREFIL, *Are We Alone? The Possibility of Extraterrestrial Civilizations.* New York, N. Y.: Charles Scribner's Sons, 1981.

SAGAN, C., "On the Origin and Planetary Distribution of Life," *Radiation Research,* 15 (1961), 174–92.

———, "Direct Contact Among Galactic Civilizations by Relativistic Interstellar Spaceflight," *Planetary and Space Science,* 11 (1963), 485–98.

———, *Communication With Extraterrestrial Intelligence.* New York, N. Y.: Doubleday, 1973.

———, "On the Detectivity of Advanced Galactic Civilizations," *Icarus,* 19 (1973), 350–52.

———, "The Solar System," *Scientific American,* 233, no. 3 (1975), 23–31.

———, *The Dragons of Eden: Speculations on the Evolution of Human Intelligence.* New York, N. Y.: Random House, 1977.

———, "Eavesdropping on Galactic Civilizations," *Science,* 202 (1978), 374–76.

———, *Broca's Brain: Reflections on the Romance of Science.* New York, N. Y.: Random House, 1979.

———, *Cosmos.* New York, N. Y.: Random House, 1980.

SAGAN, C., and F. DRAKE, "The Search for Extraterrestrial Intelligence," *Scientific American,* 232, no. 5 (1975), 80–89.

SAGAN, C., and W. I. NEWMAN, "The Solipsist Approach to Extraterrestrial Intelligence," *Quarterly Journal of the Royal Astronomical Society,* 24 (1983), 113–21.

SAGAN, C., and T. PAGE, Eds., *UFOs—A Scientific Debate.* New York, N. Y.: W. W. Norton, 1972.

SAGAN, C., L. S. SAGAN, and F. DRAKE, "A Message From Earth," *Science,* 175 (25 February 1972), 881–84.

SAGAN, C., and R. G. WALKER, "The Infrared Detectability of Dyson Civilizations," *The Astrophysical Journal,* 144 (1966), 1216–18.

SAUNDERS, M. W., "Databank for an Inhabited Extrasolar Planet: Purpose, Indication, and Installation," *Journal of the British Interplanetary Society,* 30 (1971), 349–58.

SCHOPF, J. W., "The Evolution of the Earliest Cells," *Scientific American,* 239, no. 3 (1978), 110–38.

SCHUBERT, G., and C. COVEY, "The Atmosphere of Venus," *Scientific American,* 245, no. 1 (1981), 66–74.

SCHUBERT, G., D. J. STEVENSON, and K. ELLSWORTH, "Internal Structure of the Galilean Satellites," *Icarus,* 47 (1981), 46–59.

SCHWARTZMAN, D. W., "The Absence of Extraterrestrials on Earth and the Prospects for CETI," *Icarus,* 32 (1977), 473–75.

SHEAFFER, R., "Project Ozma II," *Spaceflight,* 17 (1975), 421–23.

———, "NASA Contemplates Radio Search for Extraterrestrial Intelligence," *Spaceflight,* 18 (1976), 343–47.

———, "1977 SETI Progress Report," *Spaceflight,* 19 (1977), 307–10.

SHERROD, CLAY P. with THOMAS L. KOED, *A Complete Manual of Amateur Astronomy.* Englewood Cliffs, N. J.: Prentice-Hall, Inc., 1981.

SHKLOVSKII, I. S., and C. SAGAN, *Intelligent Life in the Universe.* New York, N. Y.: Dell, 1966.

SIEVER, R., "The Earth," *Scientific American,* 233, no. 3 (1975), 83–90.

SIMPSON, G. G., "The Nonprevalence of Humanoids," *Science,* 143 (1964), 769–75.

SINGER, C. E., "Galactic Extraterrestrial Intelligence I—The Constraint on Search Strategies Imposed by the Possibility of Interstellar Travel," *Journal of the British Interplanetary Society,* 35 (1982), 99–115.

SMITH, A. G., "Constraints Limiting the Rate of Human Expansion into The Galaxy," *Journal of the British Interplanetary Society,* 34 (1981), 363–66.

———, "Settlers and Metals—Industrial Supplies in a Barren Planetary System," *Journal of the British Interplanetary Society,* 35 (1982), 209–17.

SMITH, B. A., and others, "Encounter with Saturn: Voyager 1 Imaging Science Results," *Science,* 212 (10 April 1981), 163–91.

SODERBLOM, L. A., "The Galilean Moons of Jupiter," *Scientific American,* 242, no. 1 (1979), 88–100.

SQUYRES, S. W., and others, "Liquid Water and Active Resurfacing on Europa," *Nature,* 301 (1983), 225–26.

STAFF AT NATIONAL ASTRONOMY AND IONOSPHERE CENTER, "The Arecibo Message of November 1974," *Icarus,* 26 (1975), 462–64.

STARR, C., "Energy and Power," *Scientific American,* 225, no. 3 (1971), 36–49.

STEPHENSON, D. G., "Factors Limiting the Interaction Between Twentieth-Century Man and Interstellar Cultures," *Journal of the British Interplanetary Society,* 30 (1977), 105–8.

———, "Extraterrestrial Cultures Within the Solar System?" *Quarterly Journal of the Royal Astronomical Society,* 20 (1979), 422–24.

———, "A Classification of Extraterrestrial Cultures," *Journal of the British Interplanetary Society,* 34 (1981), 486–90.

STEVENSON, D. J., "Models of the Earth's Core," *Science,* 214 (1981), 611–19.

STOKER, H. S., S. L. SEAGER, and R. L. CAPENER, *Energy: From Source to Use.* Glenview, Ill.: Scott, Foresman, and Co., 1975.

STROM, R. G., "Mercury: A Post-Mariner 10 Assessment," *Space Science Reviews,* 24 (1979), 3–70.

SULLIVAN, W., *We Are Not Alone,* New York, N. Y.: New American Library, 1964.

——, "What If We Succeed?" *Cosmic Search,* 1, no. 1 (1979), 37–39.

SULLIVAN, W. T., S. BROWN, and C. WETHERILL, "Eavesdropping: The Radio Signature of the Earth," *Science,* 199 (1978), 377–88.

TANG, T. B., "Fermi Paradox and CETI," *Journal of the British Interplanetary Society,* 35 (1982), 236–40.

TARTER, J., and others, "A High-Sensitivity Search for Extraterrestrial Intelligence at 18 cm," *Icarus,* 42 (1980), 136–44.

TARTER, J. C., and F. P. ISRAEL, "A Symbiotic Approach to SETI Observations: Use of Maps From the Westerbork Synthesis Radio Telescope," *Acta Astronautica,* 9 (1982), 415–19.

TAUBE, M., "Future of the Terrestrial Civilization over a Period of Billions of Years (Red Giant and Earth Shift)," *Journal of the British Interplanetary Society,* 35 (1982), 219–25.

TAYLOR, S. R., "Structure and Evolution of the Moon," *Nature,* 281 (1979), 105–10.

TIPLER, F. J., "Extraterrestrial Intelligent Beings Do Not Exist," *Quarterly Journal of the Royal Astronomical Society,* 21 (1980), 267–81.

———, "Additional Remarks on Extraterrestrial Intelligence," *Quarterly Journal of the Royal Astronomical Society,* 22 (1981), 279–92.

TOVMASYAN, G. M., ed. *Extraterrestrial Civilizations.* Erevan, USSR: Isdatel'stvo Akademii Nauk Armyanskoi SSR, 1965. Translated by Israel Program for Scientific Translations, 1967.

TOWE, K. M., "Early Precambrian Oxygen: A Case Against Photosynthesis," *Nature,* 274 (1978), 657–61.

TOWNES, C. H., and others, "The Center of the Galaxy," *Nature,* 301 (1983), 661–66.

TRAFTON, L., "Does Pluto Have a Substantial Atmosphere?" *Icarus,* 44 (1980), 53–61.

TROITSKY, V. S., L. N. BONDAR, and A. M. STARODUBTSEV, "The Search for Sporadic Radio Emission from Space," *Soviet Physics-Uspekhi,* 17 (1975), 607–9. Translated from *Uspekhi Fizicheskikh Nauk,* 113 (1974), 718–20.

TROITSKY, V. S., and others, "Search for Monochromatic 927 MHz Radio Emission from Nearby Stars," *Soviet Astronomy (AJ),* 15 (1971), 508–10. Translated from *Astronomicheskii Zhurnal,* 48 (1971), 645–47.

TURCO, R. P., and others, "Nuclear Winter: Global Consequences of Mutliple Nuclear Explosions," *Science,* 222 (1983), 1283–92.

TURK, A., and others, *Environmental Science,* Philadelphia, Pa.: W. P. Saunders, 1978.

VALENTINE, J. W., "The Evolution of Multicellular Plants and Animals," *Scientific American,* 239, no. 3 (1978), 141–58.

VAN DE KAMP, P., "Unseen Astrometric Companions of Stars," *Annual Reviews of Astronomy and Astrophysics,* 13 (1975), 295–333.

VERSCHUUR, G. L., "A Search for Narrow Band 21-cm Wavelength Signals from Ten Nearby Stars," *Icarus,* 19 (1973), 329–40.

VIEWING, D. R. J., and C. J. HORSWELL, "Is Catastrophe Possible?" *Journal of the British Interplanetary Society,* 31 (1978), 209–16.

VON DANIKEN, E., *Chariots of the Gods?,* London, England: Souvenir Press, 1969.

VON HOERNER, S., "The Search for Signals from Other Civilizations," *Science,* 134 (1961), 1839–43.

———, "The General Limits of Space Travel," *Science,* 137 (1962), 18–23.

———, "Astronomical Aspects of Interstellar Communication," *Astronautica Acta,* 18 (1973), 421–30.

———, "Population Explosion and Interstellar Expansion," *Journal of the British Interplanetary Society,* 28 (1975), 691–712.

———, "Where is Everybody?" *Naturwissenschaften,* 65 (1978), 553–57.

WALD, G., "The Origins of Life," *Proceedings of the National Academy of Sciences,* 52 (1964), 595–611.

WALKER, J. C. G., "The Search for Signals From Extraterrestrial Civilizations," *Nature,* 241 (1973), 379–81.

WASHBURN, S. L., "Evolution of Man," *Scientific American,* 239, no. 3 (1978), 194–207.

Wood, J. A., "The Moon," *Scientific American,* 233, no. 3 (1975), 93–102.

Woodwell, G. M., "The Carbon Dioxide Question," *Scientific American,* 238, no. 1 (1978), 34–43.

Wyatt, S. P., *Principles of Astronomy.* Boston, Mass.: Allyn and Bacon, 1977.

Wyllie, P. J., "The Earth's Mantle," *Scientific American,* 232, no. 3 (1975), 50–63.

Yokoo, H., and T. Oshima, "Is Bacteriophage ϕX174 DNA A Message From an Extraterrestrial Intelligence?," *Icarus,* 38 (1979), 148–53.

Young, A., and L. Young, "Venus," *Scientific American,* 233, no. 3 (1975), 71–78.

Zuckerman, B., and J. Tarter, "Microwave Searches in the USA and Canada," in *Strategies for the Search for Life in the Universe,* ed. M. D. Papagiannis. Dordrecht, Holland: D. Reidel Publishing Co., 1980.

INDEX